International Studies in the History of Mathematics and its Teaching

The International Studies in the History of Mathematics and its Teaching Series creates a platform for international collaboration in the exploration of the social history of mathematics education and its connections with the development of mathematics. The series offers broad perspectives on mathematics research and education, including contributions relating to the history of mathematics and mathematics education at all levels of study, school education, college education, mathematics teacher education, the development of research mathematics, the role of mathematicians in mathematics education, mathematics teachers' associations and periodicals.
The series seeks to inform mathematics educators, mathematicians, and historians about the political, social, and cultural constraints and achievements that influenced the development of mathematics and mathematics education. In so doing, it aims to overcome disconnected national cultural and social histories and establish common cross-cultural themes within the development of mathematics and mathematics instruction. However, at the core of these various perspectives, the question of how to best improve mathematics teaching and learning always remains the focal issue informing the series.

More information about this series at http://www.springer.com/series/15781

Gert Schubring
Editor

Interfaces between Mathematical Practices and Mathematical Education

Springer

Editor
Gert Schubring
Department of Mathematics
Federal University of Rio de Janeiro
Rio de Janeiro, Brazil

Institut für Didaktik der Mathematik
Bielefeld University
Bielefeld, Germany

ISSN 2524-8022 ISSN 2524-8030 (electronic)
International Studies in the History of Mathematics and its Teaching
ISBN 978-3-030-01616-6 ISBN 978-3-030-01617-3 (eBook)
https://doi.org/10.1007/978-3-030-01617-3

Library of Congress Control Number: 2018965212

This Springer imprint is published by the registered company Springer Nature Switzerland AG
The registered company address is: Gewerbestrasse 11, 6330 Cham, Switzerland

Introduction

This volume is a result of discussions first launched by two papers published in the French journal, *Revue d'Histoire des Mathématiques*, by Bruno Belhoste (1998) and myself (Schubring 2001). Belhoste had published a strong plea for a reassessment of the role of teaching in the history of mathematics. He criticised the abstinence of historians of mathematics in addressing this issue and researching the social and intellectual space in which the production of mathematics occurs. As he remarked, this abstinence is all the more astonishing since such approaches are even "banal" meanwhile in the historiography of science. He thought it necessary to affirm that there exists no completely autonomous sphere of theoretical production (Belhoste 1998, p. 289).

Regarding what he called the socialisation of mathematical knowledge within communities of specialists and communities of users, teaching was understood by Belhoste as a special modality of the socialisation of knowledge in which the recipient finds himself in the situation of learning. As he emphasised, teaching thus constitutes an essential component of normal science, in the sense of Thomas Kuhn. To achieve progress in this role of teaching, Belhoste proposed three major research directions. The first, where some research had been done already, should be on institutional history: the role of teaching in the organisation of the disciplinary field and the professionalisation of the mathematical community; here, he alluded to the evidence of "un monde des professeurs". A second direction should be the representations realised in teaching activities, which contribute to structuring the disciplinary field; one case here would be the variations in delimiting "elementary" and "higher" mathematics and, another, the changing notions of rigour in mathematics. The third research direction presented probably the most challenging issue: the impact of teaching activities upon the development and diffusion of mathematical practices (ibid. and passim).

Belhoste had thus aptly systematised aspects of research into the interactions between teaching and development of mathematics. The chapters in this volume present research within this now unfolding field.

In my reaction to Belhoste's paper, I highly welcomed his approach and the research programme. I underlined his criticism of "l'idée fausse que la production mathématique peut être séparée *a priori* par l'historien des conditions de sa reproduction" (Belhoste 1998, p. 298).[1] Moreover, I proposed to deepen his methodological approach. For the third research field, he had made a strong claim:

> les institutions et représentations structurant le champ disciplinaire *déterminent* en effet des pratiques, c'est-à-dire des modes de travail, qui modèlent l'activité mathématique. (ibid.; my emphasis)[2]

Although constituting a strong claim, the examples given for it—for instance, of elliptic functions, which rather reveal differences of personal style—would not convince an "internalist", who would be ready to admit a certain influence but not a "determination". I suggested that one needs a more elaborated methodological approach: the basic terms used, "production" and "reproduction", express already a separation between the two aspects; and, inevitably, such categories imply a hierarchy between invention and transmission, where production is attributed to the primary status and teaching a derived status. As such, one would not be enabled to conceive of contributions of teaching to research. The essential challenge for the historiography of mathematics is, hence, to understand mathematical invention in all its complexity (Schubring 2001, p. 297).

In that note, I outlined a conception for an interdisciplinary approach to account for this complexity, based on Niklas Luhmann's sociological systems theory of science. Thus, communication constitutes the basic activity of science. For primary communication to succeed, a common language and shared culture are necessary. From the emergence of modern states and especially since the establishment of the first systems of public education, national states have constituted the primary units of communication. Within such systems, due to socialising educational processes, young people begin to share a certain number of significations and cultural and social values and extend to shared epistemologies of science. Over extended periods, religious values constituted the basis for such shared sociocultural values. Piaget and Garcia have clearly elaborated the indissoluble connection between the epistemology of a scientific discipline and its sociocultural embedding:

> In our view, at each moment in history and in each society, there exists a dominant epistemic framework, a product of social paradigms, which in turn becomes the source of new epistemic paradigms. Once a given epistemic framework is constituted, it becomes impossible to dissociate the contribution of the social component from the one that is intrinsic to the cognitive system. That is, once it is constituted, the epistemic framework begins to act as an ideology which conditions the further development of science. (Piaget/Garcia 1989, p. 255)

The second key pattern of systems theory consists in conceiving of a society or a state as a system constituted by a plurality of subsystems that interact with each other in manners determined by their functions, which the subsystems exert in relation to the other subsystems or the system itself. To analyse mathematical

[1] The wrong idea that mathematical production can be separated a priori by the historian from the conditions of its reproduction.

[2] The institutions and representations structuring the disciplinary field in fact *determine* practices that is to say modes of work, which shape mathematical activity.

production, it is therefore necessary to investigate the functions that mathematics exerts in relation to other subsystems in a given period, culture and state (Schubring 2001, p. 302). Clearly, the educational system and, in particular, the systems of secondary schooling and higher education exert key roles. Even academies and research institutes are not autonomous systems, but subsystems related to the professionalisation of scientists.

Actually, I undertook extended research on conceptual developments in mathematics within this framework on the number conceptions underlying the development of analysis in different mathematical communities in Europe, mainly in France and Germany (Schubring 2005).

I was glad for an invitation since it gave welcome occasion to discuss this methodological conception with colleagues sharing this interest. The *Association for the Philosophy of Mathematical Practices* (APMP) is well known for its promotion of methodological reflections on mathematics. The APMP invited me to organise a workshop within its fourth meeting, from 23 to 27 October 2017, in Salvador/Bahia (Brazil)—a workshop exactly on the relations of mathematics teaching to mathematical practices. Clearly, my first invitation to participate in this workshop went to Bruno Belhoste. Unfortunately, he did not accept, telling me that he had changed entirely his scientific interests and was no longer active in this area. Even so, we had a meeting with fruitful discussions. The present volume contains contributions made at the workshop, revised and considerably extended, and contributions by researchers who for various reasons were not able to come to the APMP meeting.

The first chapter by Christine Proust relates her research on the mathematical practices of the scribes in the Old Babylonian culture. In earlier studies, she had already investigated how the masters perfected and developed the practices of arithmetic and geometry already established while teaching apprentices (see Proust 2014).[3] In this chapter, her focus is on how the fundamental knowledge learned by the later masters during their education impacted upon their mathematical production. Her sources for this investigation are quite unique and revealing as well: clay tablets of cuneiform school exercises, preserved by being reused in house constructions.

The mathematical curriculum of the scribal schools has been reconstructed and identified as organised in three levels, elementary, intermediate and advanced. Analysing the basic operations and techniques at the elementary level, she shows that the apprentices had to solve problems by relating numerical values of metrological units to their corresponding numbers in sexagesimal place value notation. The basic means for this were multiplication tables and tables of reciprocals. Exploring the duality between these two representations, the detailed reconstruction of mathematical techniques provided at the intermediary and advanced levels allowed going beyond the "linear" problems of the elementary level and the development of methods to tackle problems with plane figures and with solids.

[3] An emblematic result should be quoted here: "Developments in mathematics are the result of both the activity of teaching and interaction within a community of scholars" (Proust 2014, p. 92).

In the second chapter, Jens Høyrup investigates discourses for legitimating a mathematical procedure. To better reveal their characteristics, the first areas to be evaluated are the decidedly different cultures of the mediaeval Italian *abbacus* culture and the Old Babylonian culture. Explanations for the practices analysed might not be based on preceding arguments but on what the learner might accept as evident; he calls this the "locally obvious". The question then turns out to be when critique begins and such evidences are challenged.[4] An example from Old Babylonian mathematics provides a first instance of demonstration through critique. For the emergence of demonstration from critique, Høyrup studies the practice of the seven liberal arts and sees in the discovery of irrationality a decisive pivotal point—the elaboration of elements, from Hippocrates on, documents the growing tendency towards axiomatisation of geometry as a manner for answering critique and refuting it. He shows how axiomatisation became completed as deductivity and exemplifies this by showing how Euclid's propositions 1 to 10 of Book II can be understood as a critique of the Old Babylonian cut-and-paste practices. The later practice of commentaries is presented as turning axiomatisation into an ideology.

In the third chapter, Jorge Molina studies a key dimension for the development of mathematical practices, the cultural and epistemic values attributed to mathematics by the dominant religion in a society. Given that the Catholic faith constituted a basic cultural foundation in great parts of Europe, particularly in countries of the Roman languages, it is highly pertinent to assess the determination of the role of mathematics within the churches' appreciation of science. As it turns out, there is no uniform position, but, instead, rather different ones. Basically, two different epistemological strands were influencing the positions: Aristotelicism and Cartesianism. It was the Jesuits, educating the great majority of the Catholic youth with their enormous number of colleges in these countries, who based their conceptions on their reception of Aristoteles; there, mathematics exerted a marginal role. At the Sorbonne in Paris during the sixteenth century, Aristotle was the unquestionable authority not only in philosophy but also in physics—and Loyola studied there (see Torretti 1999, p. 2). A minority in France was the Order of the Oratorians, which not only adopted Cartesianism but was also decisive in the reform of the *Académie des Sciences* in 1699 which established the organisation as disciplinary *classes*, in particular the *classe des mathématiques* and the research task for its members (Shank 2004, p. 286). And there was the intra-Catholic reform movement of Jansenism, theologically based on Augustine, attributing a key role in achieving faith to mathematics. It was these minority strands that, eventually, together with Cartesianism, prepared the way for rationalism in the Enlightenment movement and the new social function of mathematics.

In Chap. 4, Jemma Lorenat (Claremont, Cal./USA) presents the results of an extensive and novel analysis of French geometry textbooks published in the period after the revolutionary years until the middle of the 1830s, a period of centralisation

---

[4] It is revealing that Dardi's numerical "proof" for "minus times minus makes plus", by arguing on 8 by 8 as 10 – 2 by 10 – 2 was taken up by Cardano in his *Regula Aliza* to "prove" that "minus times minus makes minus"—provoking thus later critique (Schubring 2005, p. 44).

of education and of political conservatism from 1814. Despite this, the period was very rich in new developments in geometry. While geometry had basically stagnated during the eighteenth century, an enormous productivity set in, launching new branches of geometry: geometry of position, geometry of situation, descriptive geometry, etc. Lorenat's focus is how and in which manner these new mathematical developments entered the textbooks and thus enabled their dissemination and further deepened research. In particular, the analysis focuses on pedagogical values, which shaped how new research results were integrated into geometry textbooks. Especially noteworthy is Charles Dupin's 1813 claim that scientific progress proves to become fruitful only "when it also leads to the progress of elementary treatises", thereby becoming "general knowledge" (Lorenat, this volume). Other authors, such as Claude Bergery, joined Dupin in emphasising the importance of presenting new results for practitioners. On the other hand, many of the textbooks were destined for students of the *École polytechnique*, which had then fixed prescribed study courses and was hence not receptive to include many of the new geometry topics into its curriculum.

The chapter evidences in an impressive manner the amount of novel notions in the various geometries being exposed quite rapidly in textbooks. Yet, there were apparently institutional pressures which made some authors return to earlier editions without these innovations. Lorenat refers to institutional restrictions and limitations of the mathematical background of the students. In fact, since 1814, mathematics teaching had become severely reduced in the general curriculum of the colleges. There is a section on the reception of geometry textbooks, mainly by peer reviews in journals.

Gert Schubring develops more systematically, in Chap. 5, the arguments sketched out in his comments on the paper by Belhoste. Remarkably, and in contrast to the ambiguous situation of integrating research into geometry textbooks in France, he presents two cases where German Gymnasium teachers presented decisive mathematical innovations in schoolbooks when these developments were not yet even fully elaborated—set theory—or still meeting rejections or reservations by mathematicians and philosophers—non-Euclidean geometry. The notion of "element", so emblematic for the exposition of mathematical knowledge, is analysed together with the conception of elementarising knowledge—an approach owed to the Enlightenment and its seminal project to make scientific knowledge generally accessible. There is evidence that, within this project, textbook authors were credited as contributing even to the progress of science. It might be thought that teaching mathematics within higher education—the first research direction proposed by Belhoste—necessarily and trivially impacts upon developing mathematical research; yet, such teaching might be just for general education, as it had been for over centuries, or for service courses for the formation of engineers. Thus, the cases showing that it is the function of teacher training that impacts upon the emergence of research in pure mathematics present precious specification of the functions of teaching, though cautionary remarks are important in that teaching might also exert a dogmatising effect on mathematical practices.

Tinne Hoff Kjeldsen presents, in Chap. 6, two case studies for her approach to investigating the interrelations between developments in mathematics and the conditions for these developments, addressing various concrete contexts beyond teaching, scientific as well as non-scientific. For these studies, she applies her methodology of "multiple perspective approach" developed by her for historical studies on mathematics. The two case studies focus on quite recent issues of applied mathematics, one on the impact of World War II on the development of mathematical programming and the other on attempts to develop theoretical biology by applying mathematics. Starting from a theorem in the mathematical theory of linear programming, she analyses the revealing interactions between the actors developing these theorems and the theory, especially of game theory, with various US instances and agencies for applying them in WWII and post-war contexts. The initiatives for establishing a theoretical biology reveal characteristic resistances of the biological community against applying mathematical modelling; here, too, the main context is institutions and agencies in the USA. Within her multiple perspective approach, the author shows the genuine importance of teaching and the introduction of study courses for establishing (sub-)disciplines and a related body of knowledge.

In the last chapter, Carlos Tomei comments as a working mathematician on the discussed issues of interfaces. He emphasises the need for interdisciplinary investigations to attain significant insights into these interactions. He exemplifies this importance through two cases: the effect of ideology and religion on teaching and the mutual influence of practices of different subjects and their impact upon society. He indicates various intriguing episodes from the history of mathematics, which deserve to be freshly investigated under such broader conceptual approaches.

Together, the chapters of this volume give evidence for the promising perspectives opened by this approach for interrelating two such decisive dimensions of mathematical practices—teaching and research—in interdisciplinary investigations. At the same time, they show possible avenues of future research made accessible by the contributions in this volume to achieve further insights into this hereby unfolded new framework.

Gert Schubring

References

Belhoste, Bruno. 1998. Pour une réévaluation du rôle de l'enseignement dans l'histoire des mathématiques. *Revue d'Histoire des Mathématiques* 4: 289–304.

Piaget, Jean and Rolando Garcia. 1989. *Psychogenesis and the History of Science*. New York: Columbia University Press.

Proust, Christine. 2014. Does a Master always write for his students? Some evidence from Old Babylonian Scribal Schools. In *Scientific Sources and Teaching Contexts Throughout History: Problems and Perspectives*, eds. Alain Bernard and Christine Proust, 69–94. Dordrecht, Heidelberg, London, New York: Springer.

Schubring, Gert. 2001. Production Mathématique, Enseignement et Communication. Remarques sur la note de Bruno Belhoste, "Pour une réévaluation du rôle de l'enseignement dans l'histoire des mathématiques" parue dans la RHM 4(1998), p. 289–304. *Revue d'Histoire des Mathématiques* 7: 295–305.

Schubring, Gert. 2005. *Conflicts Between Generalization, Rigor, and Intuition: Number Concepts Underlying the Development of Analysis in 17th-19th Century France and Germany*. New York: Springer.

Shank, J. B. 2004. "There was no such thing as the 'Newtonian Revolution,' and the French initiated it." Eighteenth-Century Mechanics in France before Maupertuis. *Early Science and Medicine* 9(3): 257–292.

Torretti, Roberto. 1999. *Philosophy of Physics*. Cambridge: Cambridge University Press.

Contents

About the Authors

Jens Høyrup is professor emeritus at Roskilde University, Section for Philosophy and Science Studies, and honorary research fellow, Institute for the History of Natural Science, Chinese Academy of Sciences. He is author of books and articles in the history of mathematics and other fields.

Tinne Hoff Kjeldsen is professor in the history of modern mathematics at the Department of Mathematical Sciences, University of Copenhagen, Denmark. She has published on history and philosophy of modern mathematics, history of mathematics for the learning of mathematics and history and learning of mathematical models in modern science. She has published and edited books for both specialists and for a more general public.

Jemma Lorenat teaches and researches the history of mathematics at Pitzer College in California. Her focus is on geometry in the nineteenth century, with a particular interest in techniques of visualisation. She has contributed articles to *Historia Mathematica*, *Science in Context*, *Archive for the History of Exact Sciences* and *Revue d'histoire des mathématiques* and is the editor of the Years Ago column for *Mathematical Intelligencer*.

Jorge Alberto Molina is a professor of philosophy at the State University of Rio Grande do Sul, Brazil. He is the author of essays and books devoted to the history and philosophy of mathematics and logic. He is also the author of translations from the Latin to the Portuguese and Spanish of modern philosopher's texts accompanied with commentaries.

Christine Proust is a historian of mathematics and ancient sciences, specialist of cuneiform sources. She is a member of the laboratory SPHERE (CNRS and University Paris Diderot, Paris, France) as directrice de recherche. She studied the organisation of mathematical curriculum in scribal schools during the Old Babylonian period (early second millennium BC). She has published two books on Nippur's sources: *Tablettes mathématiques de Nippur* (2007) and *Tablettes* *mathématiques de la collection Hilprecht* (2008). She is one of the editors, altogether with K. Chemla and A. Keller, of the Springer's book series *Why the Sciences of the Ancient World Matter*.

Gert Schubring is a retired member of the Institut für Didaktik der Mathematik, a research institute at Bielefeld University, and at present is a visiting professor at the Universidade Federal do Rio de Janeiro (Brazil). His research interests focus on the history of mathematics and the sciences in the 18th and 19th centuries and their systemic interrelation with social-cultural systems. One of his specialisations is history of mathematics education. He has published several books, among which is *Conflicts between Generalization, Rigor and Intuition: Number Concepts Underlying the Development of Analysis in 17th–19th Century France and Germany* (New York, 2005).

Carlos Tomei is a full professor of mathematics at Pontifícia Universidade Católica do Rio de Janeiro. He received his PhD at the Courant Institute, New York, in 1982. His interests in the human aspects of mathematics led him to participation in activities related to teaching and writing on the history and social aspects of mathematics.

Chapter 1
Foundations of Mathematics Buried in School Garbage (Southern Mesopotamia, Early Second Millennium BCE)

Christine Proust

Abstract In this chapter, I suggest an analysis of mathematical texts by relying on notions, concepts, and tools that were instilled during their early education in scholars who were active in Southern Mesopotamia during the Old Babylonian period (early second millennium BCE). The sources considered are mathematical texts from scribal schools that flourished in the Ancient Land of Sumer. I propose to examine the content of the garbage from these schools to access the "internal meanings" of advanced mathematical texts. First, I outline some characteristics of the mathematics taught in elementary education, drawing mainly from sources found in Nippur. Then, I discuss two mathematical texts in detail and show how they can be interpreted using the tools that were taught to (or invented by) their authors or users. The first example (CBS 1215) deals with the extraction of reciprocals by factorization; the second (CBS 12648) deals with volume problems using coefficients. The conclusion underlines the mathematical work that some ancient erudite scribes accomplished in order to emancipate them from the world of elementary education.

Keywords Coefficient · Cuneiform mathematics · Curriculum · Factorization · Reciprocal · Scribal schools · Sexagesimal place-value notation · Volume

1 Basic Knowledge, Fundamental Knowledge

In this chapter, I suggest a way of addressing some scholarly mathematical texts by relying on notions, concepts, and tools that were instilled in the ancient scribes during their early education. This approach could be termed as "emicist," in the sense of "approach to societies or social groups through 'internal' meanings," given to this term by Olivier de Sardan (2015, p. 65). As far as ancient mathematics is concerned,

C. Proust (✉)
CNRS and Université Paris Diderot, Paris, France
e-mail: christine.proust@uni-paris-diderot.fr

G. Schubring (ed.), *Interfaces between Mathematical Practices and Mathematical Education*, International Studies in the History of Mathematics and its Teaching,
https://doi.org/10.1007/978-3-030-01617-3_1

such a search for "internal meaning" has been brilliantly undertaken by Jens Høyrup, who developed a method of "close reading" of mathematical cuneiform texts and revealed the geometrical dimension of apparently algebraic procedures (Høyrup 2002 and Chap. 2, this volume). In this chapter, I argue that another point of access to "internal meanings" would be to consider the mathematical education of the ancient authors themselves. How the fundamental notions that scholars acquired in their early education impact their mathematical production? There is generally no hope of answering this kind of question due to the scarcity of sources and information on the ancient contexts of education. However, cuneiform sources offer a unique opportunity to lift the veil: thousands of school exercises, mainly from the city of Nippur, a cultural center situated at the Northern limit of the Ancient Land of Sumer, have come down to us. In general, school drafts were not intended to be preserved or archived, and they were usually destroyed, which explains their rarity. This is also true of the scribal schools which were active during the Old Babylonian period (early second millennium BCE) in the ancient Near East. However, a particular property of the writing support leads to the unexpected conservation of the school garbage. Once dry, tablets are almost indestructible objects and were often recycled as building materials. Part of the scribal schools' huge production was incorporated in foundations of houses, walls, floors, or fillings, and these recycling practices contributed to protecting the modest school writings against destruction. I propose to examine the content of this school garbage to access the "internal meanings" of advanced mathematical texts. In the following, I show how whole facets of the mathematical world in which the young scholars were immersed during their elementary education can be reconstructed through school texts. I argue that a modern historian can benefit from a familiarity with the basics of mathematics taught to scholars in their youth in order to penetrate the meaning of some mathematical texts.

The elementary curriculum, notably that of Nippur, is well known, and in the first part (Sect. 2), I only summarize the main results of previous studies.[1] My aim in this section is to outline some characteristics of the mathematics taught in elementary education, drawing mainly from sources found in Nippur, and to provide the reader with some basic texts, which allow an understanding of the subsequent sections. Then, I discuss some mathematical texts in detail in order to analyze how they can be interpreted using the tools that were taught to (or invented by) the authors or users of these texts. The first example deals with the extraction of reciprocals as detailed on tablet CBS 1215 (Sect. 3); the second is a third degree problem dealing with volumes as detailed on tablet CBS 12648 (Sect. 4). Both tablets are dated to the Old Babylonian period and come from the same region, the Ancient Land of Sumer in Southern Mesopotamia (the lower valley of the Tigris and Euphrates). Whatever the uncertainties about their provenience, it is quite clear that these texts, and others that are mentioned in this chapter, were produced in the context of scribal schools by authors immersed in a mathematical world similar to that developed in Nippur.

[1] For example, Veldhuis (1997), Robson (2001), and Proust (2007).

2 The Mathematical Curriculum in Nippur Schools

At the beginning of the second millennium BCE, scribal schools were particularly active in Nippur. This city has delivered abundant school archives to modern archeologists. The thousands of school drafts found by American archeologists between 1888 and 1990 provide a remarkably clear picture of teaching. Among the texts produced in the school context, Sumerian literary texts composed for educational purposes echo the content of the teaching, that is, the "scribal art" to use their own terms. For example, a literary composition called "Dialogue between two scribes" by its translators (Civil 1985; Vanstiphout 1997) summarizes the school program as follows[2]:

> Having been at school for the required period,
> I am now an expert in Sumerian, in the scribal art, in reading tablets, in calculation and in accounting.
> I can even speak Sumerian.

This extract indicates what was learnt in schools: cuneiform script, writing, reading and speaking Sumerian,[3] calculation, and accounting. Exercises written by the young scribes provide more detail on this content and on the order in which it was learnt. The curriculum in Nippur has recently been reconstructed, and it has been shown that schooling was organized in three levels, called by Veldhuis (1997) elementary, intermediate, and advanced. The mathematical curriculum in Nippur was in the same way organized in three levels (Proust 2007, Forthcoming-a). The elementary level consisted in learning metrological systems, multiplication and reciprocal tables, the sexagesimal place-value system (see explanation Sect. 2.1), and correspondences between measurement values and numbers in sexagesimal place-value notation. The intermediate level consisted in solving short problems dealing with the surfaces of squares, the calculation of reciprocals, and diverse linear situations. I do not enter into the detail of the curriculum, to which an abundant literature has already been devoted,[4] but limit myself here to presenting a selection of some school texts of elementary, intermediate, and advanced level that illuminate the main characteristics of the mathematical world to which the scholars were exposed in their early years of education.

---

[2] Below is the translation of lines 36–38 of the text "Dialogue between two scribes," reconstructed after several duplicates from Nippur and some other Southern cities (my translation after Civil and Vanstiphout).

[3] Sumerian is a dead language at this time, while the mother tongue of the young scribes attending the schools was Akkadian, a Semitic language.

[4] See Veldhuis (1997), Robson (2001), Proust (2007) and bibliography provided by these publications.

2.1 Elementary Level

The first of the mathematical entities introduced in elementary education were measurement values. All the measurement values commonly used in mathematical and economic practices were enumerated in increasing order, in lists of capacity, weight, surface and length measurements, the so-called metrological lists. Then a set of numerical tables was introduced (reciprocals, multiplication tables, square, and square root) operating on numbers in sexagesimal place-value notation (hereafter SPVN). At the same time, a system of correspondence between measurement values and numbers in sexagesimal place-value notation was established through metrological tables similar to that translated in Table 1.1.

Metrological and numerical tables provided the basic toolbox used in mathematics in scribal schools, and, for this reason, they are also indispensable for a modern reader to understand mathematical cuneiform texts. I give below the translations of the tables that will be useful for the following discussions.[5] The order adopted below is not necessarily that of the curriculum.[6]

Metrological Tables

The metrological tables of lengths establish a correspondence between length measurements and numbers in SPVN. These tables were used for linear dimensions belonging to a horizontal plan, for example, for the sides of a field or the base of a canal. The metrological table of lengths translated below (Table 1.1) is of unknown provenience.[7]

The metrological tables of heights establish a correspondence between measurements values of vertical lines and numbers in SPVN. These tables were used, for example, for the depth of a canal or the thickness of a brick in order to calculate their volume. Indeed, a volume unit is defined in the Old Babylonian metrology as a surface unit with a thickness of 1 *kuš* (ca. 50 cm), which corresponds to the number 1 in SPVN—see Appendix A. The metrological table of heights translated below (Table 1.2) comes from Nippur.[8]

[5] The same metrological and numerical tables are attested in many duplicates from Nippur and other proveniences. The variations from one source to another being insignificant, any copy can be used indifferently. The sources that I present in this section come from Nippur if possible. However, in some cases, I use sources from other proveniences that are in better condition than those from Nippur. All the intermediate level sources used in part 2.2 come from Nippur. The tablet possibly reflecting the advanced level quoted in part 2.3 comes from another Southern city, possibly Larsa. All the sources quoted in this chapter are dated to the Old Babylonian period.

[6] The order adopted in the curriculum, metrological tables of capacity, weight, surface, length and height, numerical tables of reciprocal, multiplication, and so on, is not the most intuitive for a modern reader. For an approach closer to the curriculum, see Proust (2007, Forthcoming-a).

[7] Tablet MS 3869/11 belongs to the Schøyen Collection and is of unknown provenience, probably from Southern Mesopotamia. This tablet was published in (Friberg 2007, p. 119); photo available at https://cdli.ucla.edu/P252942

[8] Tablet HS 243 is now kept at the University of Jena, and was published in (Hilprecht 1906, pl. 27; Proust 2008, n°31). The photo is available at https://cdli.ucla.edu/P388162

Table 1.1 Metrological table of lengths (MS 3869/11)[a]

Obverse		Reverse	
1 *šu-si*	10	5 *kuš*	25
2 *šu-si*	20	1/2 *ninda*	30
3 *šu-si*	30	1/2 *ninda* 1 *kuš*	35
4 *šu-si*	40	1/2 *ninda* 2 *kuš*	40
5 *šu-si*	50	1/2 *ninda* 3 *kuš*	45
6 *šu-si*	1	1/2 *ninda* 4 *kuš*	50
7 *šu-si*	1:10	1/2 *ninda* 5 *kuš*	55
8 *šu-si*	1:20	1 *ninda*	1
9 *šu-si*	1:30	1 1/2 *ninda*	1:30
1/3 *kuš*	1:40	2 *ninda*	2
1/2 *kuš*	2:30	2 1/2 *ninda*	2:30
2/3 *kuš*	3:20	3 *ninda*	3
1 *kuš*	5	3 1/2 *ninda*	3:30
1 1/3 *kuš*	6:40	4 *ninda*	4
1 1/2 *kuš*	7:30	4 1/2 *ninda*	4:30
1 2/3 *kuš*	8:20	5 *ninda*	5
2 *kuš*	10		
3 *kuš*	15		
4 *kuš*	20		

[a]The length unit šu-si (a finger) represents about 1.6 cm; the unit kuš (a cubit) is 30 šu-si, and the unit ninda is 12 *kuš* (see synthesis in Appendix A)

Table 1.2 Metrological table of heights, HS 243, from Nippur

Translation (reconstruction)

Obverse		Reverse	
1 *šu-si*	2	2 *kuš*	2
2 *šu-si*	4	3 *kuš*	3
3 *šu-si*	6	4 *kuš*	4
4 *šu-si*	8	5 *kuš*	5
5 *šu-si*	10	1/2 *ninda*	6
6 *šu-si*	12	1/2 *ninda* 1 *kuš*	7
7 *šu-si*	14	1/2 *ninda* 2 *kuš*	8
8 *šu-si*	16	1/2 *ninda* 3 *kuš*	9
9 *šu-si*	18	1/2 *ninda* 4 *kuš*	10
1/3 *kuš*	20	1/2 *ninda* 5 *kuš*	11
1/2 *kuš*	30	1 *ninda*	12
2/3 *kuš*	40		
1 *kuš*	1		

Table 1.3 Metrological table of weights, surfaces, and volumes, MS 2186

Obverse

1/2 *še* silver	10
1 *še*	20
1 1/2 *še*	30
2 *še*	40
2 1/2 *še*	50
3 *še*	1
4 *še*	1:20
5 *še*	1:40
6 *še*	2
7 *še*	2:20
8 *še*	2:40
9 *še*	3
10 *še*	3:20
11 *še*	3:40
12 *še*	4
13 *še*	4:20
14 *še*	4:40
15 *še*	5
16 *še*	5:20
17 *še*	5:40
18 *še*	6
19 *še*	6:20
20 *še*	6:40
21 *še*	7
22 *še*	7:20
23 *še*	7:40
24 *še*	8
25 *še*	8:20
26 *še*	8:40

Reverse

27 *še*	9
28 *še*	9:20
29 *še*	9:40
1/6 *gin*	10
1/6 *gin* 10 *še*	13:20
1/4 *gin*	15
1/4 *gin* 5 *še*	16:40
1/3 *gin*	20
1/3 *gin* 15 *še*	25
1/2 *gin*	30
1/2 *gin* 15 *še*	35
2/3 *gin*	40
2/3 *gin* 15 *še*	45
5/6 *gin*	50
5/6 *gin* 15 *še*	55
1 *gin*	1

The metrological tables of weights establish a correspondence between weight measurements and numbers in SPVN. The following tablet contains a section for small values, which could be used also for surfaces and volume (Table 1.3).[9]

[9]Tablet MS 2186 of the Schøyen Collection (Friberg, 2007, p. 110) is of unknown provenience, probably from Southern Mesopotamia. The photo is available online (https://cdli.ucla.edu/P250902). About the use of the section for small values of tables of weights as tables of surfaces and volume, see Proust (Forthcoming-b).

Table 1.4 Multiplication table HS 217a

Obverse		Reverse	
1	9	15	2:15
2	18	16	2:24
3	27	17	2:33
4	36	18	2:42
5	45	19	2:51
6	54	20	3
7	1:3	30	4:30
8	1:12	40	6
9	1:21	50	7:30
10	1:30	8:20 times 1	8:20
11	1:39		
12	1:48		
13	1:57		
14	2:6		

A Multiplication Table

A multiplication table (HS 217a)[10] provides an excellent example to explain the sexagesimal place-value notation (Table 1.4).

In the left-hand column of the table (see photo https://cdli.ucla.edu/P254585), we see one wedge, two wedges, and so on, nine wedges, one chevron, which represents ten wedges, and so on. The wedge represents the number one, and the chevron represents the number ten. So we recognize the sequence of numbers 1, 2, and so on, 10, 11, and so on, 19, 20, 30, 40, and 50 (converted into the modern decimal system). In the right-hand column, in front of 1 we read nine wedges, that is, the number 9, in front of 2, one chevron and eight wedges, that is, the number 18, in front of 3, the number 27, and so on. We recognize the multiplication table by 9. In front of 7, we expect 63, but we see one wedge, a space, and three wedges. It means that the "sixty" of "sixty-three" is represented by a wedge in the left position. We understand that the notation uses the base 60 and a place-value principle, exactly in the same way as we represent 63 s, that is, 1 min 3 s by the number 1:3. This is the so-called "sexagesimal place-value notation."

Let us continue the reading until the entry 20, which exhibits a feature of considerable importance in the following. In front of 20, we expect 20 times 9, that is, 180, that is, 3 times 60. Indeed, we see three wedges. But the notation on the clay does not indicate that these wedges represent sixties, and not units. The "3" in front of 20 is the same as the "3" in the third line. We see that the cuneiform sexagesimal place-value notation does not indicate the position of the units in the number. 1, 60, 1/60 are noted in the same way. The notation is "floating." The fact that numbers in

[10] Tablet HS 217a from Nippur is now kept at the University of Jena and was published in (Hilprecht 1906; Proust 2008); photo at https://cdli.ucla.edu/P254585

Table 1.5 Table of reciprocal ERM 14645

Obverse

The reciprocal of 2 is	30
Reciprocal of 3	20
Reciprocal of 4	15
Reciprocal of 5	12
Reciprocal of 6	10
Reciprocal of 8	7:30
Reciprocal of 9	6:40
Reciprocal of 10	6
Reciprocal of 12	5
Reciprocal of 15	4
Reciprocal of 16	3:45
Reciprocal of 18	3:20
Reciprocal of 20	3
Reciprocal of 24	2:30
Reciprocal of 25	2:24

Reverse

Reciprocal of 27	2:13:20
Reciprocal of 30	2
Reciprocal of 32	1:52:30
Reciprocal of 36	1:40
Reciprocal of 40	1:30
Reciprocal of 45	1:20
Reciprocal of 48	1:15
Reciprocal of 50	1:12
Reciprocal of 1:4	56:15
Reciprocal of 1:21	44:26:40

SPVN do not convey absolute quantitative information is one of the most remarkable aspects of school mathematics, even if quite puzzling for a modern observer.

To sum up, the numerical system adopted in this table, as well as in all the numerical tables produced in a school context, is:

- Sexagesimal
- Place-valued
- Floating

The following section shows how all these three properties, including the third, “floating,” play an essential role in the arithmetic developed in scribal schools.

A Table of Reciprocal (ERM 14645)

The first of the numerical tables that was learnt in scribal schools was the table of reciprocals.

Tablet ERM 14645 is an example similar to many others found in Nippur and elsewhere (Table 1.5).[11]

This table states that:

The reciprocal of 2 is 30.
The reciprocal of 3 is 20, etc.

[11] Tablet ERM 14645 is kept at the Hermitage Museum, St Petersburg (Friberg and Al-Rawi 2017, Sect. 3.5). Photo https://cdli.ucla.edu/P211991

Indeed, 2 times 30 is 60, that is 1 in floating notation; 3 times 20 is 60, that is 1, and so on. Entry 7 does not appear in the table because the reciprocal of 7 cannot be written with a finite number of digits in base 60. In modern terms, numbers which have finite reciprocals in a given base are said "regular" in this base, the other "irregular." In cuneiform texts, irregular numbers in base 60 appear as "igi nu," which means in Sumerian "no reciprocal" ("igi" means reciprocal, and "nu" is the negation).

The reciprocal is a fundamental notion in cuneiform mathematics because of the way in which the divisions are performed. Indeed, division of a number *a* by a regular number *b* is a sequence of two operations: finding the reciprocal of *b*, and multiplying it by *a*.

2.2 *Intermediate Level*

The written traces of the intermediate level of education are much less uniform and standardized than that of the elementary level. The intermediate level exercises in Nippur are diverse, but deal with few topics, the most attested being the computation of reciprocals, the evaluation of the surface of squares, and small linear problems. The examples detailed below, the first on reciprocal and the second on surface, show how intermediate level exercises are based on tables learnt in the elementary level.

An Extraction of Reciprocals

For example, the tablet translated below contains a typical example of extraction of a reciprocal.[12] Many school exercises of similar content, generally written on small round or square tablets, have been found in Nippur and elsewhere.

Ist Ni 10241

Obverse

4:26:40

Its reciprocal is 13:30

Reverse

4:26:40	9
40	1:30
13:30	

On the obverse, we read 4:26:40, its reciprocal 13:30. On the reverse, the algorithm used to extract the reciprocal is displayed. We can observe that the operations that

[12] Tablet Ist Ni 10241 from Nippur is now kept at the Archaeological Museums of Istanbul (Proust 2007); photo https://cdli.ucla.edu/P368962

produce this sequence of numbers are only multiplications and extractions of reciprocals[13]:

- The reciprocal of 6:40 is 9 (obtained using the reciprocal table)
  4:26:40 × 9 gives 40 (obtained using the multiplication table by 9)
- The reciprocal of 40 is 1:30 (obtained using the reciprocal table)
  1:30 × 9 gives 13:30 (obtained using the multiplication table by 9)

The algorithm can be summarized in modern terms as follows:

- 4:26:40 is decomposed into the product 6:40 × 40.

Thus the reciprocal of 4:26:40 is the product of the reciprocals of these factors, 1:30 × 9, that is, 13:30.

The algorithm is based on the factorization of the number the reciprocal of which is to be sought, and on the rule "the reciprocal of a product of factors is the product of the reciprocals of these factors." The mathematical world in which the algorithm works is that of numbers in SPVN on which multiplication and extraction of reciprocals are carried out. These operations do not require the determination of the place of units in numbers. Floating numbers are particularly well adapted to this multiplicative world.

The Calculation of the Surface of a Square

The young scribes had to navigate between two different types of entities, numbers in SPVN, and measurement values. A dozen exercises of similar content and layout devoted to the calculation of the surface of a square found in Nippur shed lights on this navigation between numbers and quantities. All of these exercises have the same layout: on the top left we see three numbers in SPVN, on the bottom right we see a small problem text. One of these exercises, UM 29-15-192, is based on quite simple data, and shows the process well. The numbers noted on the top left corner are 20, 20, and their product 6:40.

20
20
6:40

The statement noted on the bottom right corner is:

2 *šu-si* the side of a square
How much is the surface?
Its surface is 1/3 *še*.

The relationship between the numbers written on the top left (20 and 6:40) and the length and surface measurements written on the bottom right (2 *šu-si* and 1/3 *še*) is

---

[13] Here and elsewhere in the chapter, the reader is invited to check the calculations using MesoCalc, the Mesopotamian calculator implemented by Baptiste Mélès at http://baptiste.meles.free.fr/site/mesocalc.html

exactly that established by the metrological tables: in a metrological table of lengths such as MS 3869/11 (see Table 1.1), the measurement value 2 *šu-si* corresponds to 20, and in a metrological table of surfaces such as MS 2186, the measurement value 1/3 *še* corresponds to 6:40 (actually, an extrapolation is necessary to reach this correspondence, since the smallest entry on the table is ½ *še*). As we see, the two different types of entities are displayed on the tablet in two opposite corners, the corner for quantities (bottom right) and the corner for numbers in SPVN (top left), correlated in metrological tables. Tablets of this kind invite us to consider the relationships between numbers in SPVN and quantities as correspondences, and not as equalities.

The interpretation of this text by Neugebauer and Sachs, who published it, is grounded on the implicit assumption that the relationships between the measurement values and the numbers in SPVN are *equalities*. For example the length measurement 2 *šu-si* is considered to be *equal* to 20, and thus this latter number is understood as 0;0.20 *ninda*.

> "The meaning of the marginal numbers is clear: (2 šu-si)^2 = (0;0,20 GAR)^2 = 0;0,0,6,40 SAR" (Neugebauer and Sachs 1984, p. 248).[14]

A similar text was recently analyzed in the same manner by Robson (2008, p. 9), who translates the number "1:45" noted on the top left corner as "0;00 01 45". Indeed, understanding the number "1:45" as equal to a measurement value, here a number of *ninda,* forces one to restore the order of magnitude.

With this simple example, we see how taking in consideration the notions inculcated in the elementary level, namely the nature of the correspondence between numbers in SPVN and quantities, leads to understandings (and translations) of numbers that are quite different. In the following sections, I show how these differences have important consequences on the interpretation of some scholarly mathematical texts.

2.3 *Advanced Level*

The mathematical texts reflecting advanced education are more difficult to recognized because, unlike those of elementary and intermediate levels, the type of tablets and genre of texts are not specific to this level. A text written in the context of advanced education is often hard to distinguish from a text of pure erudition. Most of the scholarly mathematical texts were probably produced in the context of scribal schools but, as schools were also centers of culture and erudition, mathematical texts were not necessarily designed only for teaching or learning.[15] However, some texts can be interpreted with some reliability as syllabus of advanced mathematics.

---

[14] For Neugebauer and Sachs, "marginal numbers" means numbers noted on the top left; they note "GAR" the length unit *ninda*.

[15] A detailed discussion of this issue is developed in Proust (2014).

It is the case of a procedure text from Southern Mesopotamia, perhaps from the city of Larsa, which provides a list of solved problems exploring in a systematic way the methods for solving linear and quadratic problems. This text, YBC 4663, is composed of eight problems dealing with the cost of digging a trench; the parameters are the dimensions of the trench, the daily work assignment, and the worker wages.[16] For example, the first problem runs as follows:

YBC 4663, problem 1

Obverse[17]

1. A trench. 5 *ninda* is its length, 1 1/2 *ninda* (is its width), 1/2 *ninda* is its depth, 10 <*gin* > is the volume of the work assignment, 6 *še* [silver is the wage of the hired man].
2. The base, the volume, the (number) of workers and the silver are how much? You, in your procedure,
3. the length and the width cross, 7:30 it will give you.
4. 7:30 to its depth raise, 45 it will give you.
5. The reciprocal of the work assignment detach, 6 it will give you. To 45 raise, 4:30 it will give you.
6. 4:30 to the wage raise, 9 it will give you. Such is the procedure.

The first six problems are linear. Some of the parameters being known, the other parameters are asked for. The last two problems are quadratic: the length and width are to be found knowing their sum (or difference) and their product. In this text, as well as in others of the same genre (see Proust Forthcoming-a), some features are interesting for the present discussions. The data given in the statements are measurement values, but the operations prescribed in the procedures are carried out on numbers in SPVN. The correspondence between the measurement values given in the statements and the numbers in SPVN used in the procedures is that established in metrological tables. Thus, the first step in the procedure (implicit here) is the same as in the exercise on the square explained above : transform the measurement values given by the statement into numbers in SPVN using metrological tables. In the first six problems, the operations are only multiplications and divisions carried out on numbers in SPVN. The arithmetical world is the same as that of the elementary and intermediate levels, a world of numbers in SPVN on which only multiplications and divisions are performed. In the last two problems, additions and subtractions of numbers in SPVN are introduced, which marks a break with the linear world the first six problems evolve in.

[16] Tablet YBC 4663 is now kept at the Yale Babylonian Collection, Yale University. It was published in Neugebauer and Sachs (1945). The elements of interpretation proposed here are detailed in Proust (Forthcoming-a).

[17] My translation. The line numbers were added by the modern editors.

2.4 Linear Paradigm

To sum up, the mathematical world of the elementary and intermediate levels of Old Babylonian schools is based on a duality between the domain of quantities, and the domain of numbers in SPVN. In the latter, the fundamental operations are the multiplication and the extraction of reciprocals. In the school mathematical world, quantities are deduced from each other by multiplication or division. One of the basic problems introduced in the early advanced mathematical education was to calculate the area of a rectangle knowing its length and width, and the reverse problems (examples and more details in Proust Forthcoming-a). In the same line, the first problem in procedure text YBC 4663 quoted above calculates a cost knowing the task to be performed, the work assignment per day per worker, and the daily wages. Other examples are found in advanced mathematics, such as calculate the value of silver in relation to the value of gold, or calculate a profit knowing the benefit (Middeke-Conlin and Proust 2014). All intermediate level exercises and a majority of the scholarly mathematical texts illustrate this linear paradigm where quantities are deduced from each other by multiplication or division. The term used in later Akkadian texts to designate mathematics, *igi-arê*, which word for word means "reciprocal-multiplication," could refer to this linear paradigm.

Addition and subtraction belong to the world of accounting documented in the numerous models of contracts found in school archives. In this world, the mathematical entities are quantities, that is, measurement values or numbers of objects, animals, or people. The duality between quantities, on which additions and subtractions operate, and numbers in SPVN, on which multiplications and reciprocals operate, could be reflected by the use of different Sumerian terms to designate calculation and accounting, respectively "šid" and "nikkas." See for example the "Dialogue between two scribes" quoted above: "I am now an expert in [...] calculation and in accounting."

There is no place in this dual world for addition and subtraction of numbers in SPVN, since the addition and subtraction belong to the world of quantities, and numbers in SPVN do not. The introduction of addition and subtraction on numbers in SPVN is intimately related to the introduction of quadratic problems in the mathematical world, as shown, for example, in the procedure text YBC 4663 (Sect. 2.3). The addition and subtraction of numbers in SPVN appear to be an audacious invention that emancipates the calculator from the rules of the linear paradigm. This invention raises serious problems since addition and subtraction require information on the relative position of numbers, while SPVN do not convey such information.[18]

The categories which seem relevant from the point of view of mathematical practices in Old Babylonian Southern Mesopotamia are different from those that were shaped in the historiography of modern or premodern mathematics. The relevant distinctions are not between integers and fractions, or between rational and irratio-

[18] A discussion on this issue is proposed in Proust (2013).

nal, for example. The relevant distinctions are between numbers used in quantification, and numbers (in SPVN) used in multiplication, and among the latter, between regular and non-regular (igi nu) numbers. Problems are not classified according to their degree (first degree, second degree, or even higher degree), but rather according to the nature of the operations involved in the procedures: on the one hand linear problems, involving only reciprocals, multiplications and other operations which derive from them (division, square roots), and on the other hand non-linear (mainly quadratic) problems, also involving additions and subtractions.

In Sects. 3 and 4, I discuss two scholarly texts in which factorization and coefficients, two methods typical of the linear paradigm, are implemented.

3 The Role of Factorization in Calculating Reciprocals

Extracting a reciprocal was one of the basic techniques taught in the intermediate level of education in Nippur and elsewhere, as explained in Sect. 2.2. This technique is developed in a systematic way in CBS 1215, a tablet dated to the Old Babylonian period and of unknown provenience.[19] The text, which contains only numbers, is divided into six columns, three on the obverse and three on the reverse (right to left). The columns are divided into 21 boxes. The first entry is 2:5, the second 4:10, the double of 2:5, the third 8:20, the double of 4:10, and so on. In each box, the entry is the double of the previous one. Thus, the last entry, in the 21st box, is 2:5 times 2^{20} (10:6:48:53:20). In each box, the text is divided into 3 sub-columns (left, center, right), and the input of the calculations is identical to the output. For example, the text in the first box runs as follows:

2:5		12
25		2:24
28:48		1:15
36		1:40
	2:5	

The numerical relationships are similar to those observed previously on tablet Ist Ni 10241:

- The reciprocal of 5 is 12 (obtained using a reciprocal table.
- 12 times 2:5 gives 25 (obtained using a multiplication table by 12).
- The reciprocal of 25 is 2:24 (obtained using a reciprocal table).
- 2:24 times 12 gives 28:48 (obtained using a multiplication table by 12).
- 28:48 is the reciprocal of 2:5, the entry.

[19] CBS 1215 is now kept at the Museum of Archaeology and Anthropology, University of Pennsylvania, Philadelphia. It was published by Abraham Sachs (1947), who was the first historian to understand this puzzling numerical text. A detailed analysis is proposed in Proust (2012b). Photo: https://cdli.ucla.edu/P254479

As in tablet Ist Ni 10241, the operations are reciprocals and multiplications, which are carried out on numbers in SPVN. The algorithm is based on breaking down the entry, here 2:5, into two factors, here 5 and 25. The reciprocal sought, 28:48, is the product of the reciprocal of the factors. Each time a factor is detected, its reciprocal is marked in the right sub-column, which helps to execute and control the algorithm. However, unlike in Ist Ni 10241, the calculation does not stop when the reciprocal of the entry is found. It continues by applying the same process again:

- The reciprocal of 48 is 1:15.
- 28:48 is broken down into its factors 48 and 36.
- The reciprocals of 48 and 36 are 1:15 and 1:40, marked on the right.
- The product of the reciprocals, 1:15 × 1:40, gives 2:5.

As previously, this sequence of operations provides the reciprocal of 28:48, that is 2:5, the entry in the first section. To sum up, the algorithm in the first section is decomposed into two sequences:

- A direct sequence: the reciprocal of the entry 2:5 is calculated, it provides 28:48.
- A reverse sequence: the reciprocal of 28:48 is calculated, it provides the entry 2:5.

The algorithm is based on the decomposition into factors of the number the reciprocal of which is sought, and the main issue, for the practitioner, was to identify these factors. The first factor can be detected in the trailing part (the last digits of the entry),[20] and the second factor is obtained by dividing the entry by the first factor.

The entry in the eighth section of CBS 1215 is 4:26:40, the same as in the school exercise Ist Ni 10241, and the direct sequence in this section is similar to the school exercise:

- The trailing part of 4:26:40 is 6:40, the reciprocal of 6:40 is 9.
- 4:26:40 is broken down into the factors 6:40 and 40.
- The reciprocals of 6:40 and 40 are 9 and 1:30, marked on the right.
- The product of the reciprocals, 9 × 1:30, gives 13:30, the reciprocal of the entry.

The reverse sequence produces the entry:

- The trailing part of 13:30 is 30, the reciprocal of 30 is 2.
- 13:30 is broken down into the factors 30 and 27.
- The reciprocals of 30 and 27 are 2 and 2:13:20, marked on the right.
- The product of the reciprocals, 2 × 2:13:20, gives 4:26:40, the entry.

We see the same calculation that appears in the school exercise Ist Ni 10241, but with some differences. Indeed, in the school exercise, there are no clear vertical alignments; here, three sub-columns are formed, which, as noted above, facilitate the execution of the algorithm. In the school exercise, only the reciprocal is asked

[20] The term "trailing part" was coined by Friberg to designate the last digits of a sexagesimal number.

for, and only the direct sequence is noted; here, we see the direct and the reverse sequences, so that the first and the last number in the section are the same. Section 8 is not the only one to be attested on a school tablet; many other known school tablets from Nippur, Ur, Mari, and other places contain calculations of the same genre. All of these school exercises reproduce more or less the direct sequence of sections of CBS 1215. Despite their similarities, the functions of the school exercises and CBS 1215 seem to have been different. I come back to this issue below.

The final sections of CBS 1215 show another feature. For example, section 20 runs as follows:

5:3:24:26:40		[9]
45:30:40		1:30
1:8:16		3:45
4:16		3:45
16		3:45
	14:3:45	
	5[2:44]:3:45	
	1:19:6:5:37:30	
11:51:54:50:37:30		2
23:43:49:41:15		16
25:18:45		16
6:45		1:20
9		6:[40]
	8:53:20	
	2:22:13:20	
	37:55:33:20	
	2:31:42:13:20	
	5:3:24:26:40	

The entry is 5:3:24:26:40, that is 2:5 after 19 doublings. The trailing part is 6:40, reciprocal 9. The entry divided by 6:40, that is, multiplied by 9, is 45:30:40. This second factor does not belong to the table of reciprocals (it is not an elementary regular number), thus its reciprocal is unknown for the practitioner. The process is then iterated: the trailing part of 45:30:40 is 40, reciprocal 1:30. The number 45:30:40 is divided by 40, that is, multiplied by 1:30, which produces 1:8:16. This number is not an elementary regular number, the process is iterated again. The trailing part of 1:8:16 is 16, reciprocal 3:45. The number 1:8:16 is divided by 16, that is, multiplied by its reciprocal 3:45, which produces 4:16. Once more this number is not an elementary regular number, the process is iterated, which produces 16, an elementary regular number. As, at each step, the reciprocal of the trailing part has been marked on the right, the calculator has just to multiply the numbers which form the right sub-column. These multiplications form a central sub-column. The final result is 11:51:54:50:37:30, the reciprocal of the entry. The reverse sequence is similar.

This calculation shows how the algorithm can be iterated. Iterations appear in the text from section 11 onward. The iterations are facilitated by the layout of the algo-

$$\overline{c} = \overline{a+b} = \overline{a} \cdot \overline{(1+b\overline{a})}$$

Fig. 1.1 Extract from Sachs (1947, p. 222)

rithm. In the examples we have seen, the factors are numbers which appear in the reciprocal table, that is, elementary regular numbers. The decomposition into elementary regular factors is not unique, as several different regular trailing parts could be chosen as divisor. Hence the question: how was the trailing part selected? We observe that in the direct sequence, the trailing part is the largest elementary regular number. There is only one way to implement the algorithm. In modern terms, such decomposition is unique. However in the reverse sequence, there is not apparent rule for selecting the trailing part. It seems that several ways of implementing the algorithm are left open. It is perhaps meaningful to observe that only the direct sequences are attested in school texts: the algorithm is taught as unique. But the work on the algorithm in CBS 1215 suggests that their author(s) were aware of the fact that the algorithm could be implemented in several ways.

Furthermore, we have seen that the reverse sequence produces the entry, and thus, its result is known in advance. The series of entries appear as a paradigmatic set of values chosen in order to work on the algorithm itself, not to produce any new result. The reverse sequence appears then to play an essential role, that of verification of the algorithm. The purpose of this text seems to be the validation of the algorithm (choice of the entries, reverse algorithm), while the purpose of the school exercises is training in the implementation of the algorithm. The text also underlines the fact that the algorithm can be implemented in several ways, despite the fact that it is taught as unique in scribal schools. It seems to me that this text is not (or not only) a sourcebook which provides school exercises, but a systematic elaboration developed by masters or scholars in order to firmly validate an algorithm, the factorization, routinely used in elementary education and in advanced mathematics.

As in the case of the interpretation of UM 29-15-192 on the surface of a square, the understanding of the nature of numbers has an important impact on the interpretation of the text. For Abraham Sachs, the extraction of the reciprocal is based on a decomposition of the entry into sum, and not into product.[21] He represents the algorithm by the following algebraic formula (the number c, the reciprocal of which is to be extracted, is decomposed in sum $a + b$; the reciprocal of a number n is noted $\overline{n}$; Fig. 1.1):

Moreover, as Sachs considers the numbers as representing quantities, he restores the orders of magnitude with plenty of zeros, which makes the process quite cumbersome (see Fig. 1.2).

[21] This interpretation originates in a tablet, probably of later date, VAT 6505, which explains the algorithm for some of the entries found in CBS 1215. The procedure includes an addition. In my view, this addition reflects an incursion into the multiplication algorithm, and is not a step in the procedure (more discussion in Proust 2012a).

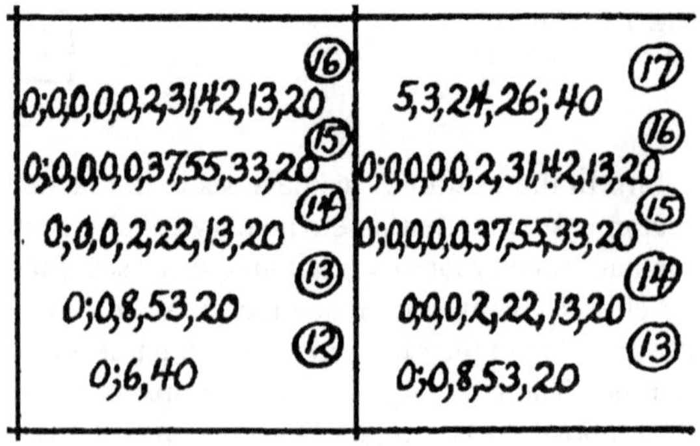

Fig. 1.2 Extract from Sachs (1947, p. 240)

4 The Role of Coefficients in Problems of Volumes

Nippur has provided archeologists with hundreds of school mathematical texts and has been by far the most prolific, among the sites of the ancient Near East, for school texts. However, very few procedure texts reflecting scholarly activities have been found in Nippur. This contrast may appear as quite surprising and may have several explanations, such as circulation of tablets during travels by scribes, chance in the excavation sites, or illegal activities of dealers. The three scholarly tablets found in Nippur, CBS 11681, CBS 19761 + Ni 5175, and CBS 12648, deal with volumes and the procedures they contain involve only multiplications and divisions.[22] Tablet CBS 11681 offers a calculation of the volume of a rectangular prism from its sides (length, width, and height) and a reverse problem, find the unknown side of a prism knowing its volume and two of its sides. Tablet CBS 19761 + Ni 5175 contains problems dealing with the volume of a pile of bricks.

In the following, I focus on CBS 12648, which illustrates interestingly one of the key tools used in the linear world, the coefficients. While the two other scholarly texts from Nippur (CBS 11681 and CBS 19761 + Ni 5175) are written in Akkadian, CBS 12648 is written in Sumerian and bears witness to the very beginning of the Old Babylonian mathematical traditions. Only the lower left part of a large multi-column tablet has come down to us. It contains several problems, only one being well preserved. Friberg (2001, pp. 149–150) was the first historian to decrypt the procedure "vaguely described in the text" (ibid., p. 150).

The preserved problem provides the volume of a rectangular prism and the ratios between its sides, asks for the sides of the prism, and gives the procedure in quite concise terms. It runs as follows:

Obverse, column *i*
Problem 2
5. 2 *še* 1/12 the volume
6. 2/3 of the length, it is the width;
7. The half of the width, it is

[22] For a complete study and bibliography, see Proust (2007), Chap. 7.

8. the depth.
9. Its length, its width
10. and its depth, how much?
11. The length, the width,
12. and the depth,
13. cross, then
14. its reciprocal detach,
15. to its volume raise.
16. The equal side (cubic root) of
17. 15:37:30
18. Is to be extracted.
19. The equal side of 15:37:30
    <is 2:30>

The procedure may appear quite mysterious at first sight because the results of the different operations which are prescribed are not specified. For example, in lines 11–13, it is prescribed to multiply the length, the width, and the depth, but the result of these operations does not appear explicitly. However, beyond the unusual style, the method used is characteristic of the mathematics of this period: it consists of transforming a problem by applying a multiplying factor so as to produce a standard problem whose resolution procedure is known.

The problem is the following: the volume (implicitly of a rectangular prism) is 2 1/12 *še* (which represents about 60 dm^3). The sides of the prism are unknown, but their ratios are given: the width is two-third of the length, and the depth is half of the width. The sides of the prism are asked for. The method consists in considering another prism of same shape as the actual prism, calculating the ratio between this reference prism and the actual prism (the volume of which is 2 1/12 *še*), and deducing the sides of the actual prism.

The sole intermediate result specified in the procedure is 15:37:30, the coefficient expressing the ratio between the reference prism and the actual prism. Finding this coefficient thus appears to be pivotal to the procedure. This number is written in SPVN, which reflects a general feature of Old Babylonian mathematical texts: all the operations prescribed in the procedures are carried out on numbers in SPVN and produce numbers in SPVN.

With these remarks in mind, the procedure can be reconstructed in detail despite its terse style (see synthesis in Table 1.6). The dimensions of the reference prism which is introduced in lines 11–14 are 1 *ninda* length, 2/3 *ninda* width (2/3 of the length), and 1/3 *ninda* depth (1/2 of 2/3 of the length). As the procedure applies to numbers in SPVN, these measurement values must be transformed into numbers in SPVN. As in other texts from the same period, these correspondences would have been provided by metrological tables:

The length 1 *ninda* corresponds to 1 according to metrological tables of length (see Table 1.1).

The width 2/3 *ninda* corresponds to 40 according to metrological tables of length (see Table 1.1).

The depth 1/3 *ninda* corresponds to 4 according to metrological tables of heights (see Table 1.2).

Table 1.6 Operations in problem 2 of CBS 12648 (in bold: data and results which appear explicitly in the text)

Lines	Text	Operations
5	2 *še* 1/12 the volume	**2 *še* 1/12** → 41:40
6–8	2/3 of the length, it is the width The half of the width, it is the depth.	1 *ninda* → 1 2/3 *ninda* → 40 1/3 *ninda* → 4
11–13	The length, the width, and the depth, cross	1 × 40 × 4 gives 2:40
14–15	Its reciprocal detach, to its volume raise	Recip(2:40) is 22:30 22:30 × 41:40 gives **15:37:30**
16–19	The equal side (cubic root) of 15:37:30 is to be extracted. The equal side of 15:37:30 < is 2:30>	The cubic root of **15:37:30** is 2:30
		1 × 2:30 gives 2:30 40 × 2:30 gives 1:40 4 × 2:30 gives 10
		Length: 2:30 → ½ *kuš* Width: 1:40 → 1/3 *kuš* Depth: 10 → 5 *šu-si*

The operation prescribed in lines 11–13 is the multiplication of the three numbers in SPVN corresponding to the length, the width, and the depth:

1 × 40 × 4 gives 2:40.

Note that these operations are prescribed indirectly by a geometrical manipulation, to "cross" the length, the width, and the depth. This produces the reference volume.

The coefficient, that is the ratio of the actual volume to the reference volume, is calculated in lines 14–15 by multiplying the former by the reciprocal of the latter. First the reciprocal of 2:40, the reference volume, is calculated, giving 22:30. The actual volume is the number in SPVN which corresponds to 2 *še* 1/12 in the metrological table for volumes, 41:40. Indeed, in such tables, for example MS 2186 (see Sect. 2.1.1, Table 1.3), 2 *še* corresponds to 40, and 1/12 *še* corresponds to 1:40, thus the measurement 2 *še* and 1/12 *še* corresponds to 41:40.[23] The coefficient of the volumes is 22:30 × 41:40, which gives 15:37:30. As noted above, this coefficient is the only intermediate result given in the procedure. The ratio of the actual volume to the reference volume being 15:37:30, the ratio of the sides of the actual prism to the sides of the reference prism is the cube root of 15:37:30 that is 2:30.

The sides of the actual prism can be calculated by multiplying the sides of the reference prism by the coefficient 2:30. However, the preserved part of the text ends by indicating that the cube root of 15:37:30 is to be calculated, but provides neither this cube root, nor the dimensions of the sides. The final steps were either omitted,

[23] The first entry in the table is 1/2 *še*; the entry 1/12 *še* is not attested but the corresponding number in SPVN is easily obtained by dividing 20 (which corresponds to 1 *še*) by 12.

or were written in the first lines of column *ii,* which is broken. I reconstruct these calculations below because the final result gives interesting information. The sides of the actual prism are obtained as follows:

Length: 1 × 2:30 is 2:30.
Width: 40 × 2:30 is 1:40.
Depth: 4 × 2:30 is 10.

The measurement values these dimensions correspond to are provided by the metrological tables (Table 1.6):

Length: 2:30 corresponds to 1/2 *kuš* (metrological table of lengths, see Table 1.1).
Width: 1:40 corresponds to 1/3 *kuš* (same table).
Depth: 10 corresponds to 5 *šu-si* (metrological table of heights, see Table 1.2).

The dimensions of the prism sought, ½ *kuš* length, 1/3 *kuš* width, and 5 *šu-si* height are that of a special brick considered as a unit of volume in several Old Babylonian mathematical texts. For example, this brick is described in a mathematical tablet from Southern Mesopotamia, YBC 4607.[24] The first section runs as follows:

A brick. 1/2 *kuš* its length,
1/3 *kuš* its width, 5 *šu-si* its height.
Its base, its volume and its equivalent capacity how much?
12 *še* and one-half a *še* its base, 2 *še* and one-12th *še* its volume,
3 1/3 *sila* 8 1/3 *gin* its equivalent capacity.

The fact that the dimensions chosen for the prism in our problem is that of a standard brick can hardly be considered as being made by chance. It reflects the circulation of mathematical practices and ideas in Southern cities of the time. This shared mathematical culture related to the quantification of space played a central role in mathematics in Old Babylonian Southern Mesopotamia (Proust Forthcoming-b). Perhaps more importantly for the present discussion, the data chosen for this problem show that the result of the procedure was known in advance. Constructing problems "backwards from the known solution" (Høyrup 2002, p. 221) is a general phenomenon in Old Babylonian mathematics; we already observed it in CBS 1215. It appears that our text CBS 12648 was not intended to produce unknown or new results, but to implement a method, that of the use of a coefficient.

Two other scholarly texts from Nippur also deal with volumes and rely on a comparison of a prism of unknown dimensions with a reference prism. This method of comparison with a reference solid was thought by Friberg to be a method of "false position" (Friberg 2001, p. 150). However, there is no technical term in these texts such as "false volume" and "true volume."[25] While from a modern point of view, there is no fundamental difference between the "false position" method and the use of

[24] This tablet, now kept at the Yale Babylonian Collection, was published by Neugebauer and Sachs (1945). For an updated study and bibliography, see Proust (Forthcoming-b).

[25] Such terms are attested in other mathematical texts. See for example Str. 368 (Thureau-Dangin 1938, p. 91–92); VAT 8391 #3 (Høyrup 2002, p. 83) the terms « truth » (gi-na = *kînum*) length and « false » (lul = *sarrum*) length. See a synthesis in Høyrup (2002, p. 100–101).

coefficients, as both consist in the same sequence of operations, it seems to me important to distinguish methods that were considered as different by the ancient actors.

The calculation itself consists only in multiplications and extractions of reciprocals carried out on numbers in floating SPVN. As before (Sect. 3), the problem is solved in the multiplicative world. No order of magnitude, no measurement unit appears in the procedure, which makes the calculation very simple, terse, and efficient. These features are destroyed by the restoration of units and orders of magnitude in interpretations, such as that of Friberg (2001, p. 150):

The ratio between the true and the false volume is
3;28,20 □ n. • n. • 9/2 = 10;25 □ n. • n. • 3/2 = 15;37,30 □ n. • n.
[Note of the present author: the notation "□ n. • n." represents a square of 1 *ninda*-side]

5 Conclusion: The Linear Paradigm and How to Get Out of It

The two examples given in Sects. 3 and 4 are emblematic of the linear paradigm, the world of *igi-arê* in which future scribes who attended scribal schools were educated and in which powerful algorithms were developed. Section 3 shows how the extractions of reciprocals are based on a method of factorization. This method was also used in other multiplicative contexts, for example for extracting square and cube roots (Proust 2012b), or for reducing rectangles in the famous tablet Plimpton 322 (Britton et al. 2011). Section 4 shows the role of coefficients in volume problems. Many other examples of the use of coefficients in solving quadratic or cubic problems can be found in Old Babylonian mathematical texts, as shown by Høyrup, who describes how "scaling and change of variable are procedures that allow the reduction of complex equations to simple standard cases" (Høyrup 2002, p. 100). The use of coefficients was a basic tool in administrative and economic transactions (Robson 1999). Standard coefficients allowed the use of metrological tables for a wide range of situations (Proust Forthcoming-b). However, the scholarly mathematical texts show how masters in scribal schools emancipated themselves from the linear paradigm in order to solve new classes of problems, and primarily quadratic problems (Høyrup 2002). A pedagogical list of problems such as YBC 4663 shows how the addition and subtraction of numbers in SPVN were introduced in the context of quadratic problems (Sect. 2.3).

It must be underlined that the scenario described in this chapter, the mathematical work in the linear paradigm and the efforts to break free from it, is based on sources from Southern Mesopotamia and dated to the first part of the Old Babylonian period, that is, before the destruction of Southern cities around the mid-seventeenth century. Different scenarios could be described from other mathematical sources, such as that of central Mesopotamia after the collapse of South and the emigration of scholars from Southern cities (Larsa, Nippur, Ur, or Uruk) to Kish or Sippar, or sources from the Old Babylonian city of Susa in Western Iran.

To fully perceive the extent of the inventiveness of scholars who gravitated around the scribal schools of the early Old Babylonian Southern Mesopotamia, it seems to me important to be aware of the prevalence of the linear paradigm. This mathematical context makes the introduction of addition and subtraction of numbers in SPVN even more remarkable in a world where these operations made no sense. This mathematical work goes unnoticed in an arithmetic world based on the four operations that we are familiar with. Its innovative and daring character is discernible only if considered from the point of view of a mind imbued with the linear paradigm.

Acknowledgements I thank the people of *La Filature du Mazel* (Valleraugue, France) who offered me a wonderful co-working space to write this chapter.

Appendix

The diagrams below represent the metrological systems and correspondences with SPVN according to Old Babylonian sources from Nippur. The arrows represent the factors between different units (e.g., *gin* ←180– *še* means 1 *gin* is equal to 180 *še*); the numbers below the units are the numbers in SPVN that correspond to these units in the metrological tables (e.g., 20 below *še* means that 1 *še* corresponds to 20 in metrological tables).

Length and other horizontal dimensions (1 *ninda* represents about 6 m)

ninda	←12–	*kuš*	←30–	*šu-si*
1		5		10

Heights and other vertical dimensions (1 *ninda* represents about 6 m)

ninda	←12–	*kuš*	←30–	*šu-si*
5		1		2

Surface and volume (a surface of 1 *sar* is that of a square of 1 *ninda*-side; a volume of 1 *sar* is that of a rectangular cuboid of 1 *sar*-base and 1 *kuš*-high)

gan	←100–	*sar*	←60–	*gin*	←180–	*še*
1:40		1		1		20

Weight (1 *mana* represents about 500 g)

mana	←60–	*gin*	←180–	*še*
1		1		20

References

Britton, John P., Christine Proust, and Steve Shnider. 2011. Plimpton 322: A review and a different perspective. *Archive for History of Exact Sciences* 65: 519–566.

Civil, Miguel. 1985. Sur les "livres d'écoliers" à l'époque paléo-babylonienne. In *Miscellanea Babylonica, Mélanges Offerts à M. Birot*, ed. Jean-Marie Durand and J.-R. Kupper, 67–78. Paris: Erc.

Friberg, Jöran. 2001. Bricks and mud in metro-mathematical cuneiform texts. In *Changing Views on Ancient Near Eastern Mathematics*, ed. Jens Høyrup and Peter Damerow, 61–154. Berlin: Dietrich Reimer Verlag (Berliner Beiträge zum Vorderen Orient).

———. 2007. *A Remarkable Collection of Babylonian Mathematical Texts*. Vol. 1. New York: Springer (Manuscripts in the Schøyen Collection: Cuneiform Texts).

Friberg, Jöran, and Farouk N.H. Al-Rawi. 2017. *New Mathematical Cuneiform Texts*. New York: Springer.

Hilprecht, Herman Vollrat. 1906. *Mathematical, Metrological and Chronological Tablets from the Temple Library of Nippur*. Philadelphia: University of Pennsylvania (Babylonian Expedition).

Høyrup, Jens. 2002. *Lengths, Widths, Surfaces. A Portrait of Old Babylonian Algebra and Its Kin*. Berlin & Londres: Springer (Studies and Sources in the History of Mathematics and Physical Sciences).

Middeke-Conlin, Robert, and Christine Proust. 2014. Interest, price, and profit: An overview of mathematical economics in YBC 4698. *Cuneiform Digital Library Journal*: 2014–2013.

Neugebauer, Otto, and Abraham J. Sachs. 1945. *Mathematical Cuneiform Texts*. New Haven: American Oriental Series & American Schools of Oriental Research.

———. 1984. Mathematical and metrological texts. *Journal of Cuneiform Studies* 36: 243–251.

Olivier de Sardan, Jean-Pierre. 2015. Emic and the Actors' point of view. In *Epistemology, Fieldwork and Anthropology*, ed. Jean-Pierre Olivier de Sardan and Antoinette Tidjani Alou, 65–82. New York: Palgrave Macmillan US.

Proust, Christine. 2007. *Tablettes Mathématiques de Nippur (With the Collaboration of Antoine Cavigneaux)*. Istanbul: Institut Français d'Études Anatoliennes, De Boccard.

———. 2008. *Tablettes Mathématiques de la Collection Hilprecht (With the Collaboration of M. Krebernik and J. Oelsner)*. Leipzig: Harrassowitz (Texte und Materialen der Frau Professor Hilprecht Collection).

———. 2012a. Guessing an algorithm beyond numbers displayed on a clay tablet: A sample for Old Babylonian period. In *Explicit Versus Tacit Knowledge in Mathematics*, ed. Tom Archibald, Jeanne Peiffer, and Norbert Schappacher, 20–23. Oberwolfach: Mathematisches Forschungsinstitut Oberwolfach (Mathematisches Forschungsinstitut Oberwolfach Report).

———. 2012b. Interpretation of reverse algorithms in several Mesopotamian texts. In *The History of Mathematical Proof in Ancient Traditions*, ed. Karine Chemla, 384–422. Cambridge: Cambridge University Press.

———. 2013. Du calcul flottant en Mésopotamie. *La Gazette des Mathématiciens* 138: 23–48.

———. 2014. Does a Master always write for his students? Some evidence from Old Babylonian Scribal schools. In *Scientific Sources and Teaching Contexts throughout History: Problems and Perspectives*, ed. Alain Bernard and Christine Proust, 69–94. Dordrecht, Heidelberg, London, New York: Springer (Boston Studies in the Philosophy of Science).

———. Forthcoming-a. How did mathematics teachers work four thousand years ago? Curricula and syllabuses in Mesopotamia. In *Resources in Mathematics Teachers' Professional Activity*, ed. Luc Trouche, Gislaine Gueudet, and Birgit Pepin. Berlin: Springer (Advances in Mathematics Education).

———. Forthcoming-b. Volume, brickage and capacity in Old Babylonian mathematical texts from Southern Mesopotamia. In *Cultures of Computation and Quantification*, ed. Karine Chemla, Agathe Keller, and Christine Proust. Heidelberg, New York: Springer (Why the Sciences of the Ancient World Matter).

Robson, Eleanor. 1999. *Mesopotamian Mathematics, 2100-1600 BC. Technical Constants in Bureaucracy and Education*. Vol. 14. Oxford: Clarendon Press (Oxford Editions of Cuneiform Texts).

———. 2001. The Tablet House: A Scribal School in old Babylonian Nippur. *Revue d'Assyriologie* 95: 39–66.

———. 2008. *Mathematics in Ancient Iraq: A Social History*. Princeton: Princeton University Press.

Sachs, Abraham J. 1947. Babylonian mathematical texts 1. *Journal of Cuneiform Studies* 1: 219–240.

Thureau-Dangin, François. 1938. Textes Mathématiques Babyloniens. Leiden: Ex Oriente Lux.

Vanstiphout, Herman L.J. 1997. School Dialogues. In *The Context of Scripture, I: Canonical Compositions from the Biblical World*, ed. William W. Hallo, 588–593. Leiden, New-York, Köln: Brill.

Veldhuis, Niek. 1997. *Elementary Education at Nippur, The Lists of Trees and Wooden Objects*. PhD Dissertation Thesis, University of Groningen, Groningen.

Chapter 2
From the Practice of Explanation to the Ideology of Demonstration: An Informal Essay

Jens Høyrup

Abstract The following discusses the practice of mathematical argument or demonstration—at first based on what I shall speak of as "the locally obvious", that is, presuppositions which the interlocutor—or, in case of writing, the imagined or "model" reader—will accept as obvious; next in its interaction with *critique*, investigation of the conditions for the validity of the seemingly obvious as well as the limits of this validity. This is done, in part through analysis of material produced within late medieval Italian *abbacus* culture, in part from a perspective offered by the Old Babylonian mathematical corpus—both sufficiently distant from what we are familiar with to make phenomena visible which in our daily life go as unnoticed as the air we breathe; that is, they allow *Verfremdung*. These tools are then applied to the development from argued procedure toward axiomatics in ancient Greece, from the mid-fifth to the mid-third century BCE. Finally is discussed the further development of ancient demonstrative mathematics, when axiomatization, at first a practice, then a norm, in the end became an ideology. The whole is rounded off by a few polemical remarks about present-day beliefs concerning the character of mathematics.

Keywords Dardi of Pisa · Explanation as demonstration · Old Babylonian mathematical critique · Critique in early Greek geometry · Hippocrates of Chios · Oinopides · Euclid · Simplicios

1 Arguing from the Locally Obvious

Let us start with this piece from Dardi of Pisa's *Aliabraa argibra*, written in 1344, presumably in Veneto.[1] It comes from the first part of the treatise, which teaches the arithmetic of monomials, binomials and polynomials containing radicals.

[1] When at home in Pisa, Dardi would obviously not be identified as coming from there. Where then did he write? The oldest manuscript (Vatican, Chigi M.VIII.170) is written in Venetian, which does

J. Høyrup (✉)
Section for Philosophy and Science Studies, Roskilde University, Roskilde, Denmark
e-mail: jensh@ruc.dk

G. Schubring (ed.), *Interfaces between Mathematical Practices and Mathematical Education*, International Studies in the History of Mathematics and its Teaching,
https://doi.org/10.1007/978-3-030-01617-3_2

The passage teaches how to divide *number* by *number plus square root,* and is based on the example $\frac{8}{3+\sqrt{4}}$. I transcribe in modern notation—Dardi has *più* where I write "+", *meno* where I have "–", and ℞ where I use $\sqrt{}$. Finally, I write the division as a fraction—this is less innocuous but useful for our later argument.[2]

Building on what he has already taught, Dardi starts by calculating that $\left(3+\sqrt{4}\right)\cdot\left(3-\sqrt{4}\right)=3^2-\left(\sqrt{4}\right)^2=5$. Then he knows (we would say that this is the definition of division, but such concepts were not Dardi's) that

$$\frac{5}{3-\sqrt{4}}=3+\sqrt{4}$$

and that

$$\frac{5}{3+\sqrt{4}}=3-\sqrt{4}.$$

What we need to find is

$$\frac{8}{3+\sqrt{4}}.$$

So far, nothing amazing. But now comes something unexpected. Dardi makes appeal to the rule of three, which tells him that

$$\frac{8}{3+\sqrt{4}}=\left(8\cdot\left[3-\sqrt{4}\right]\right)\div 5=\left(24-\sqrt{256}\right)\div 5$$

which he then in agreement with *abbacus* algebra[3] aesthetics reduces to

---

not say much. However, this manuscript uses the characteristic Venetian spelling *çenso* and the corresponding abbreviation ç. So does the Arizona manuscript, whose orthography is also northern; the last two manuscripts, written in Tuscan, still use the abbreviation ç even though writing *censo* or *cienso* when not abbreviating (actually I have not seen the Florence manuscript, but Libri's transcription of a short extract (1838, p. III, 349–359) uses "*c*," probably standing for *ç*). There is thus no reasonable doubt that the original was written in Venetian or a related dialect.

The manuscripts are discussed in Hughes (1987) and in Franci (2001, p. 3–6). I thank Van Egmond for access to his personal transcription of the Arizona manuscript.

[2] I refer to the text edition in Franci (2001, p. 59); the Chigi manuscript (fol. 12^v, original foliation; probably closer to Dardi's own text) has *m̃* instead of *meno* and *e* ("and") instead of *più* but is otherwise no different.

[3] "Abbacus" (*abbaco, abbacho*) has nothing to do with any variant of the reckoning board. It stands for practical arithmetic, but in the variant that was taught in the "abbacus school", and it calculated with Hindu-Arabic numerals on paper. Abbacus schools, existing between Genova-Milan-Venice to the north and Umbria to the south from ca 1260 to c. 1600, were frequented by artisans' and merchants' sons (also sons of patrician-merchants like the Florentine Medici) for 2 years or less around the age of 11–12.

$$4\frac{4}{5}-\sqrt{10\frac{6}{25}}.$$

What precisely was the rule of three for Dardi? Not the *problem* to find an unknown q (or p) from "if q is to p as Q to P" (where p and q may stand for "quantity" and "price", respectively), nor for *whatever method* can be used to solve that problem. The rule of three is the specific method which first multiplies and then divides, and only that. In the Italian *abbacus* school environment, it was taught in words like these:

> If some computation was said to us in which three things are proposed, then we shall multiply the thing that we want to know with the one which is not of the same (kind), and divide in the other.

This is the formulation in the Umbrian *Livero de l'abbecho* (Arrighi 1989: 9),[4] dating from around 1300; it is repeated more or less verbatim in almost all *abbacus* writings that formulate the rule—see (Høyrup 2012, p. 148–152). This is thus certainly what Dardi referred to. The rule was taught unexplained; it is indeed difficult to explain, since the intermediate product has no concrete meaning.[5]

The recourse to the rule of three was certainly meant by Dardi as an explanation. Is it a demonstration? Probably even Dardi did not think of it in terms like that, but rather as what we might express as a "reasoned procedure".

We may compare with the way we ourselves may have been taught to perform the same division—I myself around the age of 14. We would have been told to multiply the numerator and the denominator of $\frac{8}{3+\sqrt{4}}$ by $3-\sqrt{4}$,

$$\frac{8}{3+\sqrt{4}}=\frac{8\cdot\left(3-\sqrt{4}\right)}{\left(3+\sqrt{4}\right)\cdot\left(3-\sqrt{4}\right)}=\frac{8\cdot\left(3-\sqrt{4}\right)}{3^2-\sqrt{4}^2}=\frac{24-8\sqrt{4}}{9-4}=\frac{24-8\sqrt{4}}{5}.$$

Even this is a reasoned procedure, but we might spontaneously tend to see it as more akin to demonstration. But how did we know that a fraction does not change its value when numerator and denominator are multiplied by the same number? And would $3-\sqrt{4}$ be a number in the sense corresponding to the argument behind this manipulation?

Abbacus *algebra* was not taught here, but flourished from ca 1310 onward in the environment of abbacus school teachers, serving to display their competence in the competition for pupils or for municipal employment.

[4] My translation, as other translations in the following unless otherwise stated.

[5] In contrast, the two alternative methods where division precedes multiplication can be explained meaningfully: q must cost p/P as much as Q; and Q/P is as much as can be bought for one monetary unit.

It certainly was not. At an earlier moment we may have been presented with an explanation of the expansion of, say, $\frac{6}{13}$ into $\frac{5\cdot 6}{5\cdot 13}$ corresponding to this diagram:

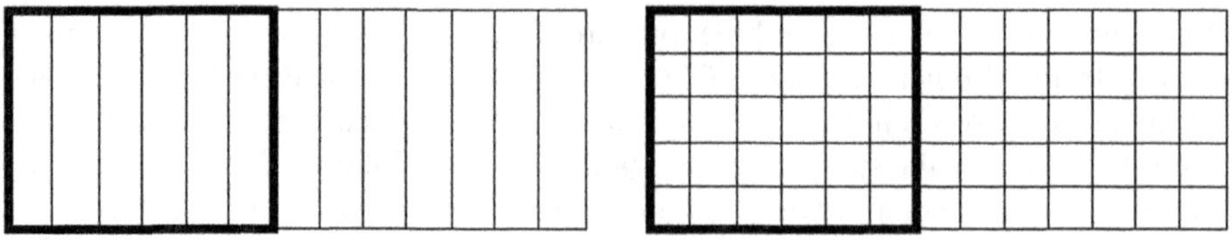

To the left, in heavy outline, we see $\frac{6}{13}$ of a rectangle—6 out of 13 equal strips. To the right the same, now 5·6 out of 5·13 equal squares, that is, $\frac{5\cdot 6}{5\cdot 13}$ of the rectangle.

To make that a rigorously valid argument in the case of irrational factors would require something like an Archimedean exhaustion. In any case, when we were confronted with $\frac{8}{3+\sqrt{4}}$ we had long forgotten the argument for the possibility of reduction or expansion of fractions (*if* we had ever been presented with one); we had just got accustomed—just as Dardi's model reader was accustomed to the use of the rule of three.

There is a difference, however, and that difference is elucidated by another passage from Dardi. Here, Dardi wants to "*prove* by a numerical example" that "minus times minus makes plus"[6]:

> I shall say, 8 times 8 makes 64, and this 8 is two less than 10, and multiply it by another 8, which is still 2 less than 10, which similarly shall make 64. This is the proof, multiply 10 times 10, it makes 100, and 10 times 2 less, it makes 20 less, and the other 10 times 2 less, it still makes 20 less, and you have 40 less, which 40 less detract from 100, remains 60. And to finish the multiplication, multiply 2 less times 2 less, which makes 4 more [*più*], join it above 60, and you have 64. And if 2 less times 2 less made 4 less, one should detract ⟨it⟩ from 60, and 56 would remain. Then it would seem that 10 less 2 times 10 less 2 would make 56, which is not true. And so it would be if 2 less times 2 less made nothing, then the multiplication of 10 less 2 times 10 less 2 would come to make 60, which is still false. So less times less by necessity needs to make plus [*più*].

This is followed in the Chigi and the Arizona manuscripts by a diagram.

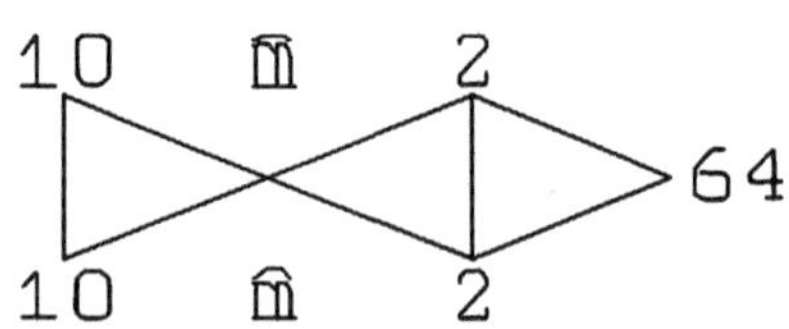

[6] Franci (2001, p. 44). The words are "dimostrare per numero" and "meno via meno fa più"—in the Chigi manuscript (fol. 5^{v}) "demostrar per numero" and "men via men fa più". Dardi distinguishes between *mostrar*, "to show", and *demostrar*, "to prove".

One may wonder at the stumbling logic in the final part of the argument—why not just *derive* that "less 2 times less 2" *must* make the 4 that has to be added to 60 in order to produce 64?[7] The other objection that might be raised—that a numerical example cannot be a proof—is easily discarded: the numerical values are just as peripheral as the actual lengths of lines entering a Euclidean proof. As Aristotle points out in *Metaphysics* M, 1078^{a}17–21 (Ross in (Aristotle, *Works*, VIII)),

> if we suppose attributes separated from their fellow-attributes and make any inquiry concerning them as such, we shall not for this reason be in error, any more than when one draws a line on the ground and calls it a foot long when it is not; for the error is not included in the premises.

As long as the argument does not depend on the actual numerical values but these just serve to carry its structure, a proof "per numero" is as good or as bad as any Euclidean demonstration by diagram.

Let us therefore concentrate on the structure. One might argue (from the meaning of multiplication as repeated addition) that adding 10 2 times less amounts to adding 20 less, and that adding 2 10 times less also amounts to adding 20 less. However, Dardi offers no argument, and in the preceding section (where number less root is multiplied by number less root, with the example $(3-\sqrt{5})\cdot(4-\sqrt{7})$ one can see that the explanation (Franci 2001, p. 43) is merely

> You shall at first multiply the numbers one by the other, that is, 3 times 4, which makes 12, and save it. And then multiply in cross the numbers times the roots, which is less, and what results is root less. Therefore multiply 3 times less root of 7, which makes root of 63, [...].

It is in the sequel that the need to multiply less by less arises. In contrast, less times more and more times less are treated as in need of no argument. They are familiar matter, just like the rule of three.

So, be it in reasoned procedure, be it in demonstration, the explanation makes use of what the learner (the presupposed or model learner) can be assumed to accept as evident—not necessarily because of preceding argument, the intuitively obvious may do as well; that is what I shall call the *locally obvious*. And, this is the crux of what precedes: *habit creates intuition* (though certainly not in one-to-one correspondence). The advice attributed to d'Alembert, "Allez en avant, et la foi vous viendra", is not too far away. Habits, on the other hand, are often linked to a *particular* practice, and thereby to the particular institutions that wield this practice. Dardi's use of the rule of three is an example, visible to us because we do not participate in abbacus school practice. Locally, it was obvious; at our distance, something that itself needs argument.

[7] Luca Pacioli, in (1494, p. I 113^{r}) actually does no better—he adds yet another possibility to be ruled out, namely that $(-2)\cdot(-2) = -2$ (Pacioli operates with negative, not just subtractive numbers), and is even more loquacious here than he usually is (to the point of being obscure).

2 Critique

What is obvious for one person (for instance, the teacher) may not be obvious to another one (for instance, the student); and what at first seems obvious may even become doubtful for the same person at second thoughts. That is where *critique* sets in, reflections about *Möglichkeit und Grenzen*, in Kant's words from the opening of the Third Critique (ed. Vorländer 1922, p. 1). I shall illustrate this with an Old Babylonian example[8]—the text YBC 6967, from somewhere between 1750 BCE and 1600 BCE. I quote the translation from Høyrup (2017, p. 45*f*).

Obv.
1. The *igibûm* over the *igûm*, 7 it goes beyond
2. *igûm* and *igibûm* what?
3. You, 7 which the *igibûm*
4. over the *igûm* goes beyond
5. to two break: 3°30′;
6. 3°30′ together with 3°30′
7. make hold: 12°15′.
8. To 12°15′ which comes up for you
9. 1` the surface join: 1`12°15′.
10. The equal of 1`12°15′ what? 8°30′.
11. 8°30′ and 8°30′, its counterpart, lay down.
Rev.
1. 3°30′, the made-hold,
2. from one tear out,
3. to one join
4. The first is 12, the second is 5
2.5. 12 is the *igibûm*, 5 is the *igûm*

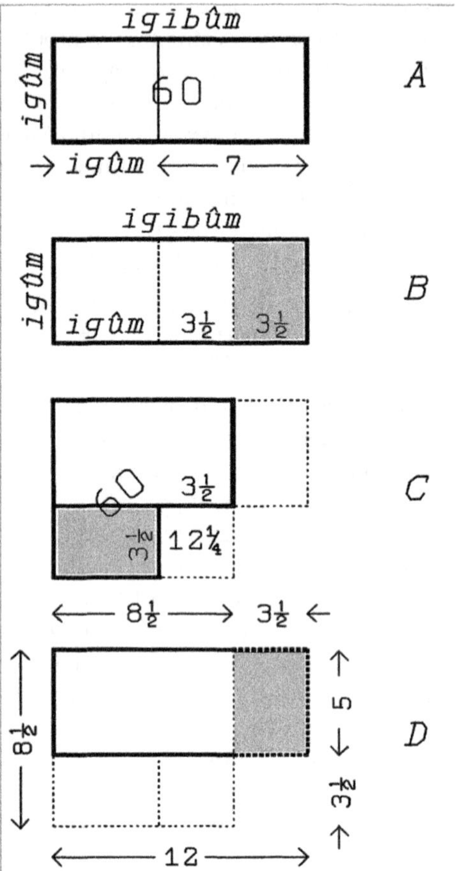

This asks for explanation—that is the price of *Verfremdung*. On the tablet, numbers are written in a floating-point place-value system with base 60. In the translation,

[8] The "Old Babylonian period" is the period 2000–1600 BCE (according to the "middle chronology"); the mathematical texts come from its second half.

they have been provided with an absolute order of magnitude. In this translation, ´ indicates decreasing and ` increasing order of magnitude; "1`12°15´" thus stands for $1\cdot60^1 + 12\cdot60^0 + 15\cdot60^{-1}$; when not needed for separation or clarity, "°" is omitted ($60^0 = 1$); "12" is thus the same as "12°".

The problem deals with two numbers belonging together in the table of reciprocals—*igûm* and *igibûm*, meaning "the reciprocal" and "its reciprocal". We should expect their product to be 1, but it is actually meant to be 1` = 60 (as mentioned, the system was floating-point). Moreover, the *igibûm* exceeds the *igûm* by 7. The words ("to make hold", "surface", "counterpart"—see imminently) show the procedure to be geometric—the two numbers are represented by the sides of a rectangle with area 1` = 60. That is shown in **A**. In **B**, the excess 7 of the *igibûm* over the *igûm* is "broken", that is, bisected—not only the segment representing it but also the appurtenant part of the rectangle, resulting in two rectangles with one side equal to the *igûm* and the other to 3°30′ = $3^1/_2$. In **C**, the outer of these rectangles is moved—the two segments of 3°30′ are "made hold", that is, arranged so that they contain a rectangle (here a square) of area 3°30´ × 3°30´ = 12°15´. To this square is joined the original rectangle transformed into a gnomon; the result is a square with area 1` + 12°15´ = 1`12°15´. Then the "equal" of this larger area is found, that is, one of the two equal sides that contain it. It is 8°30´ = $8^1/_2$. This is "laid down" together with its "counterpart"—the term may mean "to draw" or "to write", possibly also to lay down on a reckoning board (in the actual case two boards, one for each). However that may be, in **D** the "made-hold", that is, the part that was moved, is put back into its original place. Removing 3°30′ from 8°30′ leaves 5, which is the *igûm*. Putting it back yields 12, the *igûm*. We may describe the whole procedure as "cut-and-paste geometry in a square grid".

On the surface, everything here is just "seen" to be correct—but since the drawing is not found on the tablet but either just imagined by "mental geometry" or sketched on a dust board or in sand strewn on the brick-laid courtyard, even this is an instance of the locally obvious, made *obvious for us* by being transferred to our familiar medium of drawings in true proportions.

Yet one thing hides below the surface. Normally, the Babylonian reckoners, as we, would mention addition before subtraction. This is also reflected in the reversal of the order in the last two lines, 12 resulting from addition, 5 from subtraction. But in lines rev. 1–3, subtraction is performed first. The reason is regard for concrete meaningfulness: we cannot put something back in place before it is made available.

This is not evidence of "a primitive mind not yet prepared for abstraction", as has been supposed. In analogous situations, earlier texts (from around 1775 BCE) simply say "to one join, from the other tear out" (as still reflected in the order to the last two lines). At some moment, some teacher, perhaps challenged by a student, perhaps as a result of his own second thoughts, has discovered that the inherited way of speaking is deprived of concrete meaning; that is, he has engaged in *critique*.

Critique is not a conspicuous characteristic of Old Babylonian mathematical texts. I know of one other instance. Some early problems add sides of squares or rectangles to their areas without qualms, and then proceed like here, treating the segments in question as "broad lines" provided with an inherent breadth of one length unit that can be bisected.[9] Even here, later texts change their way, providing explicitly the segments with a width equal to one. Since this is done in three different ways, it seems that no less than three different teachers, each belonging to a school tradition of his own, have engaged independently in critique.

But this is all I have noticed in the Old Babylonian mathematical corpus as far as indubitable critique is concerned. After all, mathematics was basically taught as a means for administration, and even though it created a higher level of "supra-utilitarian" problems, the norms governing the practice out of which these grew asked for finding "the right number", not for theoretical justification beyond what might be didactically useful.

3 Demonstration, Critique, and the Culture of Liberal Arts

From Classical Antiquity we have the concept and ideal of "liberal arts", knowledge bodies that have no technical use but are considered goals in themselves. We may leave aside what later times did to the concept, from the Latin Middle Ages to our own world, and stick to the ancient ideal and its reality.

We should take note that the famous "cycle" of seven Liberal Arts (grammar, rhetoric, dialectic; arithmetic, geometry, astronomy, harmonics) was only formed during or after Plato's mature years, and that the supposed "seven" were normally two and nothing more—grammar, that is, good and correct use of language, and rhetoric. Augustine was no exception when he had to study on his own everything beyond these subjects (*Confessions* IV.xvi, Rouse 1912, p. 198)—but he certainly was when complaining about it. Nor was he an exception when, though an intellectually ambitious teacher of the Liberal Arts, he never had students interested in anything going beyond these matters. Things have to be reduced to due proportions.

Yet there *were*, as we know, people engaged in "liberal" mathematics during Classical Antiquity—and not only Euclid, Archimedes and Apollonios. According to Reviel Netz's estimate (1999, p. 282*f*) of the number of those who at some moment in life made a piece of explicitly reasoned mathematics, 144 have left at least minimal direct or indirect traces; perhaps some 300 were still known by name in Late Antiquity—and in total perhaps 1000, one born on the average per year, but certainly with a more uneven distribution than simple randomness would suggest (and quite possibly considerably fewer).

Their appearance also precedes the formation of the *cycle* of Liberal Arts. It *almost* had to, how could the quadrivial arts (arithmetic, geometry, astronomy and harmonics) become part of the cycle if they did not already exist? Yet we should beware that what entered the cycle were, at least by name, Pythagorean fields of interest, and to which extent these corresponded to the reasoned theoretical fields we know from Aristotle's time onward can be disputed. Even the nature of the mathematics which

[9] This notion of "broad lines" and its rather widespread occurrence is discussed in Høyrup (1995).

according to Plato should be taught to the guardians of his republic (*Republic* VII, 525D5–E3) is subject to doubt—cf. (Mendell 2008). There is no compelling evidence (if we do not count as such much later Neoplatonic interpretations) that his "arithmetic" was something like the theory of *Elements* VII–IX—after all, the word basically means "counting", and how far this meaning was stretched by Plato is not clear from his text. To Henry Mendell's arguments we may add a passage from Aristotle's *Metaphysics* N, 1090^b27–29 (which no longer concerns the state of affairs at the moment when the *Republic* dialogue is supposed to have taken place but Plato's own teaching at the moment when Aristotle was working at the Academy or later). After other objections against Plato's identification of numbers with ideas it is pointed out (trans. Ross in [Aristotle, *Works*, VIII]) that

> not even is any theorem true of them, unless we want to change the objects of mathematics and invent doctrines of our own.

That is: whatever Plato maintains in his mature philosophy about number has nothing to do with the theoretical arithmetic that had been created no later than the fourth century BCE.[10]

In any case we know that some kind of theoretical mathematics existed during the second half of the fifth century BCE. Famously, the possibility of *incommensurability* had been discovered by then, most likely by Pythagorean *mathematikoi*, and we know that first Theodoros and later Theaitetos worked on this topic—Plato's dialogue *Theaetetus*, though written after 370 BCE, can be considered testimony. From the reports about and fragments from Archytas (Diels 1951, p. I, 429–438), we also know about investigation of the three main mathematical *means* (arithmetic, geometric, harmonic). At least irrationality is beyond what could be of interest in any productive or administrative practice; a connection between the theory of means and the theory of harmonics can be presupposed, but then the theory of harmonics was a mathematical theory, and its relation to practised music questionable (questioned indeed by Aristoxenos). The theory of means was also linked to the search for two mean proportionals, which Archytas treated; even this was of no interest for administrators or master builders.

We are ignorant, however, not only of the precise arguments used by Theodoros and Archytas but also of their overall argumentative style. As regards Meton and Euctemon, we are even worse off concerning the kind of mathematical argument (if any) they used together with their astronomical observations; in any case we cannot

---

[10] A number of attempts have been made to save Plato by proving that Aristotle does not understand him—see, for instance, Tarán (1978) with references to others sharing his view, or the list in Cherniss (1944, p. X). Such attempts are misguided, what Aristotle does is to point out that the numbers Plato speaks about have nothing to do with what others mean by number—no more, indeed, than the "self-moving number" which the Pythagoreans identify with the soul (*De anima* 408^b32f).

A different question, which however does not concern us here, is whether the traces we have of Plato's views can be given a coherent and historically possible interpretation. Beyond the discussion and references in Tarán and Cherniss, see Mendell (2008, p. 128 n. 3).

say that their work belonged under the heading *theory*, the calendar was certainly a practical concern.

Happily, we know more about Hippocrates of Chios. We have his investigation of the lunules as rendered via Eudemos by Simplicios (Thomas 1939, p. I, 239–253). It is obviously reasoned—the three "classical problems", one of which (the squaring of the circle) is the inspiring background to Hippocrates's question, only make sense as theoretical problems. But there is no trace of axiomatics, the argument makes use of two principal tools, together with some properties of his diagrams which he tacitly takes for granted as intuitively obvious. One tool is that the square on the hypotenuse of a right-angled triangle equals the sum of the squares on the legs of the right angle—the "Pythagorean theorem"; the other is that the area of a circle is proportional to the square on the diameter.[11] Both had been staple knowledge for Near Eastern surveyor scribes at least since the Old Babylonian period—both are indeed used in mathematical problems from that epoch, and the proportionality of the areas of similar figures to the square of a characteristic linear dimension (side of a square, perimeter or diameter of a circle) is the fundament for the geometric part of the tables of technical constants. So, Hippocrates may have made use (*systematic* use, which is where he differs from for example Dardi) of the locally obvious); to believe that he must have known or produced a proof, for instance for the proportionality of the circular area to the square on the diameter is a *petitio principii*, proving that Greek geometry already had the character we know from the third century BCE from the tacit assumption that it had.

Further, we have Eudemos's ascription to Hippocrates of a first collection of elements—an ascription we know from Proclos's *Commentary on Book I of the Elements* 66 (Morrow 1970, p. 54). This collection is likely to have been connected to Hippocrates's teaching in Athens. The direct evidence for such teaching is a reference in Aristotle's *Meteorology* to "those around Hippocrates and his disciple Aischylos".[12] The members of this circle cannot have been engaged in practical mathematics: firstly, then they would have had no need for a collection of elements: secondly, Aristotle speaks about their opinions concerning comets. So, this earliest almost direct reference to teaching of geometry also shows it to have been teaching of geometry as a "liberal" subject.

[11] Thomas Heath (1921, p. I, 201) argues from Hippocrates's text that he knew what was to become propositions III.20–22, 26–29 and 31 in Euclid's *Elements*. This would not be amazing, they can be derived from the equality of the angles at the basis of an isosceles triangle by means of the same kind of counting as Hippocrates wields when applying the Pythagorean theorem. But it is equally possible—not least because Hippocrates makes use of these properties of figures without noticing that an argument might be needed—that he made use of what could "be seen" without having recourse to formulated propositions.

[12] Bekker (1831, p. 342^b36–343^a1). "Those around" was the standard way to refer to the circle of those who studied with a philosopher or similar teacher. Strangely, the Loeb as well as the Ross translation omits "those around", even though the Loeb edition conserves it in its Greek text. The secondary literature on the other hand (including myself on earlier occasions) has spoken about Hippocrates's teaching without questioning it.

We have no direct evidence concerning the possible teaching of Oinopides, also from Chios and slightly older than Hippocrates—at most the suggestion of Paul Tannery (1887, p. 109) that Hippocrates learned from him. Relying on Eudemos, however, Theon of Smyrna (Dupuis 1892, p. 320*f*) states that Oinopides discovered the obliquity of the ecliptic. That the planets do not move on the celestial equator was too obvious to be a discovery, so two interpretations of this passage are possible: Oinopides may have discovered that the motions of the planets not only run through a specific sequence of celestial signs (that is how matters were seen by Babylonian mathematical astronomers) but describe a great circle (which the Babylonians could not think, not possessing the notion of the heavenly vault as a sphere or hemisphere); or he may have measured the obliquity of the ecliptic (which is however so easy to do once the idea of an oblique great circle is conceived that it can hardly count as an independent discovery[13]). Our present point is a scene depicted in Plato's *Erastae* (Lamb 1927, p. 312*f*), set in the later fifth century BCE. It portrays two boys in "the grammar school of the teacher Dionysios" eagerly discussing an astronomical problem "either about Anaxagoras or about Oinopides" involving the obliquity of the ecliptic. This school (also Plato's own school according to Diogenes Laërtios (Hicks 1925, p. I, 278*f*)) was a school for "the young men who are accounted the most comely in form and of distinguished family" (thus *Erastae*), not one teaching banausic trades; here, things like Oinopides's astronomy were thus taught at least at a level that allowed eager discussion.

A different kind of evidence comes from Aristotle's writings.[14] The ideal organization of a field of knowledge as prescribed in the *Posterior Analytic* is obviously inspired by geometry[15]—not just reasoned geometry but axiomatic geometry. During the century or so that had passed since Hippocrates wrote his elements, many things could of course have changed, and Aristotle presents much material elucidating the process.

Quite a few of Euclid's definitions (or alternatives referred to by commentators) were known to Aristotle. I shall mention only two examples. Firstly, *Topica* 143^b11*f* refers to those who define the line as a "length without breadth", μῆκος ἀπλατές, exactly Euclid's definition I.1.[16] Secondly, though paraphrased and contracted, the definition of geometrical similarity referred to in *Analytica posteriora* 99^a13*f* is obviously the one offered in *Elements* VI.[17]

Definitions had been a concern in Greek philosophy for quite some time. According to Aristotle's *Metaphysics* 987^b3, (trans. Ross in [Aristotle, *Works*, VIII]), "Socrates [...] fixed thought for the first time on definitions". Whether he

---

[13] All that is needed is to measure the culmination of the sun at summer and winter solstice and to halve the difference.

[14] From Plato's dialogues, too. But they are often (already, and perhaps mainly, because of the half-poetic genre) too ambiguous to be of much use in the present discussion.

[15] "Inspired", not copying, already for the reason that Aristotelian syllogistic logic does not fit the way geometric proofs are argued. But also for other reasons, cf. McKirahan 1992, p. 135–143.

[16] Respectively Bekker 1831, p. I, 143 and Heiberg 1883, p. I, 2.

[17] Respectively Bekker 1831, p. I, 99 and Heiberg 1883, p. II, 72.

was really the first or inspired by contemporary mathematicians is probably not to be decided—not least because Aristotle speaks of ὁρισμοί but Euclid (and plausibly geometers before him) of Ὅροι, which rather means "delimitations". Aristotle is likely to have been aware that the difference was more than just a choice between synonyms.

Among Euclid's common notions, the third ("if equals be subtracted from equals, the remainders are equal" (Heath 1926, p. I, 223) is Aristotle's paradigm for an axiom or "peculiar truth" valid within a particular genus. It serves as such in *Analytica posteriora* 76ᵃ41, and again in 76ᵇ20*f*, but also in *Analytica priora* 41ᵇ21*f* as an example of a presupposition that has to be made explicit in order to avoid a *petitio principii*.

Further, Aristotle knew Euclid's second postulate—that can be seen in *Physica* 207ᵇ29–31 (Hardie & Gaye in [Aristotle, *Works*, II]):

> [mathematicians] do not need the infinite and do not use it. They postulate only that the finite straight line may be produced as far as they wish.

Euclid (Heath 1926, p. I, 154) requests (that is the meaning of "postulate") that it be possible "to produce a finite straight line continuously in a straight line".

As far as I know, the other postulates are not quoted (nor paraphrased) in the Aristotelian corpus; one, moreover, is absent where it would have been adequate to mention it, namely in *Analytica priora* 65ᵃ4–7 (Jenkinson in [Aristotle, *Works*, I]). This passage refers, as an example of hidden circular reasoning, to

> those persons [...] who suppose that they are constructing parallel straight lines: for they fail to see that they are assuming facts which it is impossible to demonstrate unless the parallels exist.

Postulate 5 (Heath 1926, p. I, 155),

> if a straight line falling on two straight lines make the interior angles on the same side less than two right angles, the two straight lines, if produced indefinitely, meet on that side on which are the angles less than the two right angles

was obviously meant to repair that calamity. Actually, it only does so halfway. It excludes hyperbolic but not elliptic geometries (precisely those where parallels do not exist). For this purpose, one has to presuppose, for example, that two straight lines cannot enclose a space, which some geometers indeed added as an axiom according to Proclos, *Commentary on Book I of the Elements* 183 (Morrow 1970, p. 143), and which in fact is used in a dubious passage in the proof of *Elements* I.4 (Heath 1926, p. I, 248, cf. p. 249) (apparently a scholion that has crept into the text).

What can we derive from these observations? In general that geometry as known to Aristotle was already striving for axiomatization—no wonder, we know from Eudemos as quoted by Proclos (*Commentary* 67, [Morrow 1970, p. 56]) that at least Theudios made a new, better arranged collection of elements, and that a number of mathematicians worked together at the Academy in Plato's time. But we also see that the enterprise had not yet led to the goal, at least not as a social undertaking—those who undertook to construct parallel straight lines while presupposing unconsciously that such lines exist were still building their reasoning on the locally obvious—and so was even Euclid in many cases, for example when he took it for granted that two lines cannot enclose a space (not to speak of his many topological intuitions).

We may also have a look at Euclid's postulate 4 (Heath 1926, p. 155), "That all right angles are equal to each other". For us, this is locally obvious—"of course, they are all 90°". Apparently, it was just as obvious until the mid-fifth century BCE—and for a similar reason. Then, according to Proclos (*Commentary* 283, [Morrow 1970, p. 220*f*]),[18] Oinopides introduced the *construction* of the perpendicular by means of ruler and compass, calling the perpendicular a line drawn "gnomonwise"—implying that until then it had been made by means of a set square (γνώμων), in which case the equality seems obvious. However, with the new construction arose the need for a *definition* of what a right angle is. In Euclid we find this (*Elements* I, def. 10, Heath 1926, p. I, 153):

> When a straight line set up on a straight line makes the adjacent angles equal to one another, each of the equal angles is right, and the straight line standing on the other is called a perpendicular to that on which it stands.

This seems to solve the problem, now we know what a right angle is, much better than the Old Babylonian surveyor-scribes (and probably the surveyor-scribes of the mid-first millennium), whose field plans show them to have distinguished between "wrong" and "right" angles, the former—those which are evidently skew—being irrelevant for area calculation and the latter—right for practical purposes—essential; see for example Høyrup (2002, p. 228). But it creates a new problem: Now it is no longer obvious that all right angles are equal, and that is needed in many proofs.

The preceding three paragraphs lapsed into old-style historiography of mathematics, which tended to forget that mathematical knowledge and practice do not exist per se but have social carriers—or, if mentioning persons, would take it for granted that these, as "mathematicians", would think "like mathematicians". The reader who had no objections will recognize how easily this lapse occurs.

Yet a problem is only one if it is a problem *for somebody*, and it only becomes a problem in the encounter with that somebody. Here, we may return to the boys from *Erastae*. If they could discuss eagerly about Oinopides and his work on the obliquity of the ecliptic, they might also challenge their teacher, and ask (this was shortly after Oinopides introduced his construction) *what* this right angle is *in itself* which he constructs (apart from being supposedly useful in astronomy, as Proclos says Oinopides had thought). The answer would be something like the Euclidean definition. And then, at a later moment, similar eager students might discover that with this definition, the equality of right angles is no longer obvious. This is *critique*, born as an endeavour from the character of the environment.

We may further remember that the environment of philosophers (to which we may count the theory-oriented mathematicians teaching elite youth just as did other philosophers) did not strive for truth in peaceful collaboration but in competition and strife. Here, *critique* would coincide with *criticism* or *challenge* of colleague-competitors.

[18] Cf. also von Fritz 1937, p. 2265 *f*.

4 Axiomatization

Critique had been a driving force in the axiomatization of geometry—axiomatization *as a goal* had not been imaginable when Oinopides and Hippocrates made their work. Not only was axiomatization the outcome of a process yet in their future; so was the discovery of the *idea* of axiomatization as a possibility. Plato's reproach to geometricians in the *Republic* (533C–D, trans. (Shorey 1930, p. II, 203), that they are

> dreaming about being, but the clear waking vision of it is impossible for them as long as they leave the assumptions which they employ undisturbed and cannot give any account of them. For where the starting-point is something that the reasoner does not know, and the conclusion and all that intervenes is a tissue of things not really known, what possibility is there that assent in such cases can ever be converted into true knowledge or science?

—this reproach may look as if Plato had observed the strivings of contemporary mathematicians to achieve axiomatic order (even though the "assumptions"/ Øπόθεσις he speaks about may also be local, as in Hippocrates's text). Whether an axiomatic structure or just locally coherent argument is meant, Plato does not accept such geometry as more than a mere mental exercise preparing the best souls for the study of dialectics,

> the only process of inquiry that advances in this manner, doing away with hypotheses, up to the first principle itself in order to find confirmation there,

which first principle is insight into "the good", no formulated axiom; and dialectic as imagined by Plato is in consequence no axiomatic system.[19]

Aristotle understood that this was a pipe dream, and that explicit axiomatization is the maximum that can be achieved. This is indeed pointed out in the very first sentences of his *Analytica posteriora* ($71^{a}1$–9, Mure in [Aristotle, *Works*, I]):

> All instruction given or received by way of argument proceeds from pre-existent knowledge. This becomes evident upon a survey of all the species of such instruction. The mathematical sciences and all other speculative disciplines are acquired in this way, and so are the two forms of dialectical reasoning, syllogistic and inductive; for each of these latter makes use of old knowledge to impart new, the syllogism assuming an audience that accepts its premisses, induction exhibiting the universal as implicit in the clearly known particular.

As we notice, this is in itself an instance of inductive dialectic as here explained.

In spite of the ambiguity of Plato's polemics (which we need not reproach him, his discourse has other concerns), these words together with the rest of the *Analytica posteriora* leave no doubt that in the mid-fourth century BCE not only Aristotle but also the geometers were familiar with the axiomatic ideal. From now on, it provided a possible format when new fields were taken up and did not need to be the unplanned outcome of a process driven by other forces. As we know, this format was to be used for example by Archimedes.

---

[19] One can argue from certain Platonic texts—but this would lead us astray—that this insight in "the good" is achieved via mystical experience. As a hint, observe the force of the images of *light*.

Critique, as argued above, had been a motive force in the process ending up in axiomatization before this process could be driven by a recognized aim. But critique was more than that. A look at *Elements* II.6 (Heath 1926, p. I, 385) will illustrate it:

> If a straight line be bisected and a straight line be added to it in a straight line, the rectangle contained by the whole with the added straight line and the added straight line together with the square on the half is equal to the square on the straight line made up of the half and the added straight line.

Whoever encounters these lines for the first time is likely to ask why this seemingly abstruse theorem is interesting. However, if we look at the diagram that accompanies the proof we recognize a familiar situation (I follow Heath, but emphasize some lines and weaken another one for clarity of the argument). Here, the bisected line is represented by *AB* and the added line by *BD*. *DM*, perpendicular to *AD*, equals *BD*. *AB* is thus the excess of length over width in the rectangle *ADMK*. If we identify *AD* with the *igibûm* and *DM* with the *igûm*, we are back at the Old Babylonian problem discussed above, and *AB* must be 7.

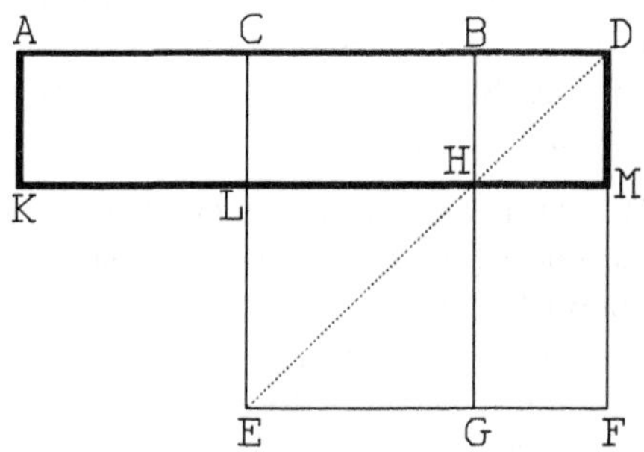

There are differences, however. Firstly, Euclid does not solve a problem: if we impress algebraic categories on his text, then he presents us with an identity. This identity can of course be used to solve problems by taking some of the magnitudes involved to be known and others unknown (for example, taking *AB* to be 7 and the area *ADMK* to be 60 will allow us to find *AD* and *BD*).

Secondly, Euclid does not move segments or areas around. At first he *constructs* the square *CDFE* on *CD*, which ensures that the angle *ADM* is really a right angle. He then draws the diagonal *DE*, which has no place in the Old Babylonian procedure. He then draws the line *BG* parallel to *CE* or *DF;* through the intersection *H* of *BG* and *DE* he draws *KM* parallel to *AB* or *EF*, and through *A* the line *AK* parallel to *CL* or *DM*. That allows Euclid to show that rectangle *ACLK* is equal to the rectangle *HMFG*, and thus that the rectangle *ADMK* equals the gnomon *CDFGHL*, whence finally the equality claimed in the enunciation. Nothing is cut, moved around and pasted, all is proved to the best standards of theoretical geometry as these had been shaped in the late fifth and the fourth centuries BCE. The proposition thus functions as a critique of the cut-and-paste procedure by which the problem was traditionally solved, showing why and under which precisely stated conditions it works—thus saving it instead of rejecting it as Plato did when he reproached geometers their "talk of squaring and applying and adding and the like" (*Republic* 527B, (Shorey 1930, p. II, 171)).

That it was also *meant* as critique and saving appears to follow from analysis of the whole sequence *Elements* II.1–10. A discussion in depth would lead too far, but see Høyrup (2002, p. 400–402). A blunt summary goes like this:

- All 10 propositions correspond in the way just sketched for II.6 to riddles or basic cut-and-paste-tricks belonging at least since ca 1800 BCE to an environment of surveyors—riddles which once inspired the Old Babylonian scribe school, but have also left their traces in a variety of written mathematical cultures until the Late Middle Ages, including Greek pseudo-Heronic practical geometry (and were therefore certainly known to Greek theoretical geometers);
- propositions 4–7 are used later in the *Elements*, mainly in Book X, the others not[20]; like many of the definitions of Book I that are never used afterwards, they represent something familiar that has to be saved for its own sake;
- propositions 2 and 3 are special cases of proposition 1; propositions 4 and 7 are different formulations of what is practically the same matter; the same can be said about propositions 5 and 6 and about propositions 9 and 10. None the less, all are proved independently, as if not only the results but also the traditional methods had to be saved through critique.

So, between Aristotle's and Euclid's times, deductivity completed as axiomatization established itself as the norm for how mathematics should be made—obviously only within the tiny group which we, like Netz, would normally accept as "mathematicians". Most of those who went through the normal syllabus of Liberal Arts would not care about anything beyond rhetoric, as pointed out above—and within that minority which had greater ambitions, most would stop at knowing a few concepts and enunciations and not care for demonstrating. That is clear from the relative popularity of Nicomachos's writings, from handbooks like those of Martianus Capella and Cassiodorus, and from Theon of Smyrna's explanation of the mathematics needed for the study of Plato. Among those who calculated or constructed for administrative or productive purposes, the norm never took root, at most we find arguments from the locally obvious—to see this, we may look at Vitruvius and the pseudo-Heronic writings.

In Euclid's time already, the effects of the "liberal" curiosity of the fifth century BCE had subsided and been replaced by institutionalized norms. For that reason, the importance of critique as a partner and root of axiomatization seems also to have subsided (after all, the critique in *Elements* II is almost certainly borrowed from late fifth or early fourth-century predecessors, as the proportion theory in *Elements* V is supposed to be borrowed from Eudoxos). Heron's *Metrica* may to some extent be considered a rewriting of practical geometry *vom höheren Standpunkt aus*—but only to a quite limited extent in a way that allows us to speak of critique.

[20] Cf. Mueller 1981, p. 301.

5 And Then?

Not too long after Euclid's third century BCE, Greek mathematics entered the age of commentaries or, in Reviel Netz's terms (1998), of "deuteronomic texts" (a somewhat broader category, encompassing also epitomes, etc.). In Simplicios's presentation of the Hippocratic fragment (early sixth century CE), he states (Thomas 1939, p. 237) that

> I shall set out what Eudemus wrote word for word, adding only for the sake of clearness a few things taken from Euclid's *Elements* on account of the summary style of Eudemus, who set out his proofs in abridged form in conformity with the ancient practice.

That illustrates a partial change of norms. Commentaries fill out and explain; at times they also discuss. Even though Simplicios is engaged in a commentary to Aristotle, he follows the commentator habits and norms even here, but mainly by filling out and, implicitly, explaining. "Adding [...] a few things from Euclid's *Elements*" means that Simplicios inserts the Hippocratic text in the axiomatized framework.

In its own way, the addition of commentaries and the standardized structuring of mathematical texts (Netz 1998, p. 268–270) is a new level of critique, arguing now *why and in which sense* the classical text that is commented upon is right and conformable to norms. But since this classical text has somewhat sacred status, this critique is uncritical—quite different from the critical critique of fellow-philosophers or teachers in the fifth to fourth centuries BCE.[21]

It is hardly necessary to point out that norms only govern practice to some extent; many causes—conflicting norms, incompetence, personal conflicting interest, and so forth—make actors deviate from them. Eudemos's lack of reference to the propositions which Simplicios feels he needs to insert may be due to fidelity to his source—he is writing a history of geometry and may have written more like a historian than as a mathematician. But it may also reflect that axiomatization in his time was still a developing *practice* and not yet fully effective as a norm. In Simplicios's age of deuteronomic texts, in contrast, the norm had become so explicit that we may see it as an *ideology*, an inextricable amalgam of the descriptive and the prescriptive, of "is" and "ought to". That ideology is still with us, admittedly more effective when governing the writing of textbooks (deuteronomic, indeed) than in mathematical research.

This ideology not only amalgamates the descriptive and the prescriptive levels. It also corresponds to the interpretation of ideology as "false consciousness". Most obviously, it disregards informatics, quantitatively the major part of twenty-first century mathematics. Already in 1970, a textbook from that field declared (quoted from the reprint (Acton 1990, p. xvii)) that

---

[21] Genuinely critical stances had not disappeared—but they had become external, attacking the whole undertaking, not trying to save or to find the "possibility and limits" of mathematical knowledge. The best example is probably Sextus Empiricus (Bury 1933, p. IV, 244–321). This is harsh but informative and informed criticism—but not critique.

> It is a commonplace that numerical processes that are efficient usually cannot be proven to converge, while those amenable to proof are inefficient [...]. The best demonstration of convergence is convergence itself.

This was written at a moment when the students using the book were supposed to work in FORTRAN, PL/1 or ALGOL—when programming was thus still transparent compared to what we find today. Every time your computer screen freezes, remember that the reason is probably an unpredicted conflict somewhere on the path from machine code through compiler to operating system or application—thus proof that the software has not been derived axiomatically from first principles. The role of beta-versions is to locate the conflicts ("bugs") that are most likely to occur—but this "critique through practice" never succeeds in doing more. The demonstrations of algorithm design remain local.

Even if we try to save the honour of mathematics by excluding informatics, the ideology misrepresents reality. In 1545, Cardano's *Ars magna* was printed. Then, gradually, the power of the tools offered by Descartes' *Géométrie* (1637) (also in analysis of the infinite and the infinitely small) was revealed. First, this transformed fundamentally what *algebra* could be; soon it also changed the global character of mathematics. Until the late nineteenth century, this whole process was founded (when not on controlled guess, as often happened) on arguments and demonstrations of no more than "local" validity, that is, premises that it seemed reasonable to accept or at least to try, but which were not built on clearly formulated first principles. Critique gradually improved the situation (even this was an epoch of competing scholars), but only the late nineteenth century was once again able to reshape mathematics on an axiomatic footing.[22]

In its merger of description and prescription, the ideology of thorough demonstration and demonstrability thus becomes false consciousness. The prescriptive aspect not only imposes a particular interpretation of the facts on the description—that probably cannot be avoided. It distorts it in a way that is easily looked through *if only one wants to*.

Recently in Italy, a nun when told by physicians that her supposed stomach ache were birth pangs, exclaimed "it is not possible, I am a nun!" Her false consciousness cannot have survived the next few hours. In general, false consciousness survives on Darwinian conditions: in some way it has to be useful. The one we have looked at here provides mathematics (that is, the mathematical establishment) with a comforting self-image, which can be projected (while the inconvenient baby, informatics, is given into adoption). It also serves to ostracize mathematical cultures that deviate from what the ideology prescribes and what we therefore claim *de*scribes

[22] These sweeping statements go beyond what can be documented in a few footnotes. But see Stedall (2010) for the development of algebra from Cardano to the early nineteenth century. Høyrup (2015, p. 29–33) covers an often overlooked aspect of the shaping and gradual reception of a Cartesian tool (the algebraic parenthesis). The painful advance in the foundation of infinitesimal calculus has been amply discussed; see, for example, Boyer (1949), Bottazzini (1986) and Spalt (2015)—not to speak of the innumerable publications dealing with particular aspects or figures.

our mathematics; thereby it serves a more direct and more indisputably political "projection of power".

References

Acton, Forman S. 1990. *Numerical Methods That Work*. Washington, DC: Mathematical Association of America.

Aristotle *Works*. Translated into English under the Editorship of W. D. Ross. 12 vols. Oxford: The Clarendon Press, 1908–1952.

Arrighi, Gino. 1989. Maestro Umbro (sec. XIII), *Livero de l'abbecho*. (Cod. 2404 della Biblioteca Riccardiana di Firenze). *Bollettino della Deputazione di Storia Patria per l'Umbria* 86: 5–140.

Bekker, Immanuel (ed.). 1831. *Aristoteles Graece*. 2 vols. Berlin: Reimer.

Bottazzini, Umberto. 1986. *The 'Higher Calculus': A History of Real and Complex Analysis from Euler to Weierstrass*. New York: Springer.

Boyer, Carl B. 1949. *The Concepts of the Calculus: A Critical and Historical Discussion of the Derivative and the Integral*. New York: Columbia University Press.

Bury, R. G. (ed., trans.). 1933. *Sextus Empiricus*. 4 vols (Loeb Classical Library). London/New York/Cambridge, MA: Heinemann/Putnam/Harvard University Press, 1933–1949.

Cardano, Girolamo. 1545. *Artis Magnae Sive de Regulis Algebraicis, Liber Unus*. Nürnberg: Petreius.

Cherniss, Harold. 1944. *Aristotle's Criticism of Plato and the Academy*. Vol. 1. Baltimore: Johns Hopkins University Press.

Descartes, René. 1637. *Discours de la Methode Pour Bien Conduire sa Raison, & Chercher la Verité dans les Sciences. Plus La Dioptrique. Les Meteores. Et La Geometrie*. Leiden: Ian Maire (imp.).

Diels, Hermann. 1951. *Die Fragmente der Vorsokratiker, Griechisch und Deutsch*. Herausgegeben von Walther Kranz. 3 vols. 6 Auflage. Berlin: Weidmann, 1951–1952.

Dupuis, Jean. (ed., trans.). 1892. *Théon de Smyrne, Philosophe Platonicien, Exposition des Connaissances Mathématiques Utiles Pour la Lecture de Platon*. Paris: Hachette.

Franci, Raffaella, ed. 2001. *Maestro Dardi, Aliabraa argibra, dal manoscritto I.VII.17 della Biblioteca Comunale di Siena. (Quaderni del Centro Studi della Matematica Medioevale, 26)*. Siena: Università degli Studi di Siena.

Heath, Thomas L. 1921. *A History of Greek Mathematics*. 2 vols. Oxford: The Clarendon Press.

———. (ed., trans.). 1926. *The Thirteen Books of Euclid's Elements*. 2nd revised edition. 3 vols. Cambridge/New York: Cambridge University Press/Macmillan.

Heiberg, Johan Ludvig. (ed., trans.). 1883. *Euclidis Elementa*. 5 vols. Leipzig: Teubner, 1883–1888.

Hicks, R. D. (ed., trans.). 1925. *Diogenes Laertius, Lives of Eminent Philosophers*. 2 vols. (Loeb Classical Library). London/New York: Heinemann/Putnam.

Høyrup, Jens, 1995. Linee larghe. Un'ambiguità geometrica dimenticata. *Bollettino di Storia delle Scienze Matematiche* 15: 3–14. English translation http://www.akira.ruc.dk/~jensh/Publications/2010{K}03_Broad_lines.PDF. Accessed 3 May 2018.

———. 2002. *Lengths, Widths, Surfaces: A Portrait of Old Babylonian Algebra and Its Kin*. New York: Springer.

———. 2012. Sanskrit-Prakrit interaction in elementary mathematics as reflected in Arabic and Italian formulations of the rule of three—and something more on the rule elsewhere. *Ganita Bhārati* 34: 144–172.

———. 2015. Embedding: Another case of stumbling progress. *Physics* 50: 1–38.

———. 2017. *Algebra in Cuneiform: Introduction to an Old Babylonian Geometrical Technique*. Berlin: Edition Open Access.

Hughes, Barnabas B. 1987. An early 15th-century algebra codex: A description. *Historia Mathematica* 14: 167–172.

Lamb, W. R. M. (ed., trans.). 1927. *Plato, Charmides. Alcibiades I and II. Hipparchus. The Lovers. Theages. Minos. Epinomis. (Loeb Classical Library)*. London/New York: Heinemann/Putnam.

Libri, Guillaume. 1838. *Histoire des Mathématiques en Italie*. 4 vols. Paris: Jules Renouard, 1838–1841.

McKirahan, Richard D., Jr. 1992. *Principles and Proofs: Aristotle's Theory of Demonstrative Science*. Princeton: Princeton University Press.

Mendell, Henry. 2008. Plato by the numbers. In *Logos and Language: Essays in Honour of Julius Moravcsik*, ed. Dagfinn Follesdal and John Woods, 125–160. London: College Publications.

Morrow, Glenn R. (ed., trans.). 1970. *Proclus, A Commentary on the First Book of Euclid's Elements*. Translated with Introduction and Notes. Princeton, NJ: Princeton University Press.

Mueller, Ian. 1981. *Philosophy of Mathematics and Deductive Structure in Euclid's Elements*. Cambridge, MA/London: MIT Press.

Netz, Reviel. 1998. Deuteronomic texts: Late antiquity and the history of mathematics. *Revue d'Histoire des Mathématiques* 4: 261–288.

———. 1999. *The Shaping of Deduction in Greek Mathematics: A Study in Cognitive History*. Cambridge: Cambridge University Press.

Pacioli, Luca. 1494. *Summa de Arithmetica Geometria Proportioni et Proportionalita*. Venezia: Paganino de Paganini.

Rouse, W. H. D. (ed.). 1912. *Augustine, Confessions*. 2 vols. (Loeb Classical Library). Trans. W. Watts. London/New York: Heinemann/Macmillan.

Shorey, Paul. (ed., trans.). 1930. *Plato, The Republic*. 2 vols. (Loeb Classical Library). London/New York: Heinemann/Putnam, 1935.

Spalt, Detlev D. 2015. *Die Analyse im Wandel und im Widerstreit; Eine Formierungsgeschichte ihrer Grundgeschichte*. Freiburg & München: Karl Alber.

Stedall, Jacqueline. 2010. *From Cardano's Great Art to Lagrange's Reflections: Filling a Gap in the History of Algebra*. Zürich: European Mathematical Society.

Tannery, Paul. 1887. *La géométrie grecque. Comment son histoire nous est parvenue et ce que nous en savons. Essai critique. Première partie, Histoire générale de la géométrie élémentaire*. Paris: Gauthiers-Villars.

Tarán, Leonardo. 1978. Aristotle's classification of number in Metaphysics M 6, 1080a15–37. *Greek, Roman and Byzantine Studies* 19: 83–90.

Thomas, Ivor (ed., trans.). 1939. *Selections Illustrating the History of Greek Mathematics*. 2 vols. (Loeb Classical Library). London/New York: Heinemann/Putnam. 1939–1941.

von Fritz, Kurt. 1937. "Oinopides", col. 2258–2272. In *Paulys Real-Encyclopädie der Classischen Altertumswissenschaften*, ed. Neue Bearbeitung begonnen von Georg Wissowa, vol. 17, 2. Stuttgart: Alfred Druckenmüller.

Vorländer, Karl. (ed.). 1922. *Immanuel Kant, Kritik der Urteilskraft*. Fünfte Auflage. Leipzig: Felix Meiner.

Chapter 3
Catholicism and Mathematics in the Early Modernity

Jorge Alberto Molina

Abstract This text discusses certain aspects of the efforts made by several clergymen of the Roman Catholic Church to produce, disseminate and teach mathematics, especially during the seventeenth century. Firstly we examine the place of geometry in the erudite culture of the time. Secondly we discuss the question of whether the efforts of these priests received strong support from Catholic institutions. Thirdly we analyse why these priests believed that mathematics was useful both for the defence and propagation of the Catholic faith. Finally we examine their philosophical commitments, specially their attitudes towards Aristotelianism and Cartesianism.

Keywords History of mathematics · Mathematics, culture and society · Philosophy of mathematics · Roman Catholic Church and science · Science in the early modern period

1 The Place of Mathematics in Seventeenth Century Culture

The phenomenon of Catholicism's contribution to Mathematics is well known and has attracted the attention of science historians and philosophers since the pioneering works of Henri Bosmans and later, in the 1950, of François de Dainville, on mathematics and the Jesuits. In the early modern period several Roman Catholic clergymen and theologians, not all of them Jesuits, showed, a strong interest in mathematics. Some of them were renowned mathematicians, others philosophers or authors of textbooks on mathematical sciences that helped disseminate knowledge related to the field. Examples include the following: Jesuit Christopher Clavius (1538–1612), Jesuit André Tacquet (1612–1660), Jesuit Gregoire de Saint Vincent (1584–1667), Jesuit Paul Guldin (1577–1643), Minim Marin Mersenne (1588–1648), Jansenist Antoine Arnauld (1612–1694), Oratorian Nicolas Malebranche (1638–1715), Oratorian Bernard Lamy (1640–1715), Oratorian Charles René

J. A. Molina (✉)
Universidade Estadual de Rio Grande do Sul, Porto Alegre, RS, Brazil
e-mail: jorge-molina@uergs.edu.br

G. Schubring (ed.), *Interfaces between Mathematical Practices and Mathematical Education*, International Studies in the History of Mathematics and its Teaching,
https://doi.org/10.1007/978-3-030-01617-3_3

Reynaud (1656–1728), Jesuat Boaventura Cavalieri (1598–1647),[1] Jesuit Giovanni Saccheri (1667–1733) and Pierre Varignon, a priest at the Saint Ouen parish in Caen (1654–1722), all of whom made contributions to mathematics on different levels. Clavius, Cavalieri, Guldin, Gregoire de Saint Vincent and perhaps Saccheri are now considered creative mathematicians, who advanced our knowledge of mathematics. Clavius became known as the author of the most famous Latin edition of Euclid's *Elements* which contained commentaries and corrections to the proofs given by the Greek geometer. Cavalieri was the author of one of the most well-known methods for determining the ratio between the areas of two curvilinear figures: the method of indivisibles.[2] Guldin is known as the author of *Centrobarycae*, whose first book deals with the determination of the centre of gravity of bodies and whose second book contains that which is known as Guldin's rules, used for calculating the areas and volumes of figures generated by the rotation of a line around an axis. Arnauld became known because of his *Nouveaux Élémens de Géométrie*, a successful handbook on elementary geometry in which he attempted to explain plane geometry in a different way than presented by Euclid's *Elements*. The Oratorians,[3] Malebranche, Reyneau and Lamy played very important roles in introducing the infinitesimal calculus in France.[4] Jean Prestet, a disciple of Malebranche and also an Oratorian, was the author of another well-known textbook *Élémens des mathématiques ou principes généraux de toutes les sciences qui ont les grandeurs pour objet* (1675), and it is difficult to determine what in the content of the work came from Prestet and what came from Malebranche.[5] Mersenne, although not a creative mathematician, made an important contribution to the development of this science by posing several mathematical questions whose solutions required efforts that were significant enough to advance the mathematical understanding of the time.[6]

[1] The Jesuati (Jesuates) were a religious order founded by Giovanni Colombini of Siena in 1360. The order was abolished by Pope Clement IX in 1668.

[2] Cavalieri gave two presentations of his theory: one of them is known as the collective method, the other one as the distributive method. See Andersen (1985).

[3] The Congregation of the Oratory of Jesus and Mary was a Roman Catholic Society of apostolic life of Catholic clergymen founded in 1611 in Paris by Pierre de Bérulle. They were very interested in education and founded several schools (*collèges*) in France. On the teaching of mathematics in the Oratorians schools see Costabel (1964) and Belhoste (1993).

[4] On the group of Oratorians united around Malebranche see Robinet (1960b). Malebranche, the most famous member of the Oratorians, changed his mind about the nature of mathematics. At first, he was greatly influenced by Descartes and believed that algebra was the most important area of mathematics. But afterwards he became familiar with the infinitesimal calculus developed by Leibniz and was one of his supporters (Robinet 1961).

[5] This work by Prestet shows what we can consider it the initial phase of Malebranche's mathematical thought. See Robinet 1960a.

[6] About Mersenne Pascal said: "Il avait un talent tout particulier pour former de belles questions; en quoi il n' avait peut-être pas de semblable; mais encore qu'il n'eût pas un pareil bonheur à les resoudre, et que ce soit proprement en ceci que consiste tout l' honneur, il est vrai néanmoins qu'on lui a obligation, et qu'il a donné l' occassion de plusieurs belles découvertes, qui peut-être n'auraient jamais été faites s'il n'y eut excité les savants". Among the questions raised by Mersenne are the following: Are mathematics useful to theology and physics? Is the quadrature of the circle possible? What are conic sections useful for? (Taussig 2018).

In his *La verité des sciences contre les Sceptiques ou Pyrrhoniens*, Mersenne has also given us a description of early seventeenth century views on mathematics. According to Mersenne, mathematics deals with abstract finite quantities. These quantities may be continuous, and are then the object of geometry, or discrete, and in this case are the object of arithmetic. Arithmetic gave birth to logistic, the art of calculus, and to algebra. These were considered to belong to the category of pure mathematics. The category of applied mathematics embraced optics, mechanics, pneumatics, hydraulic, astronomy and harmony. There was disagreement among philosophers regarding the origin of mathematical entities. Some accepted the Aristotelian position, which asserts that, as Mersenne argued, mathematical entities were obtained from the objects that we are able to perceive with our senses. On the other hand, those philosophers influenced by the *Commentary of Proclus on Euclid's Elements* maintained that they were innate.[7] These different views referring to the origin of mathematical entities lead us to two methods of defining the first mathematical objects. We can begin with the bodies and abstract from them their surfaces to obtain the notions of surface and plane surface. The border of a surface is a line, a continuous magnitude without depth and breadth. The extreme element of a line is a point. Contrarily we can go the other way beginning with the points, whose composition and union originates lines. The composition of lines gives birth to the surfaces. A surface with depth is a solid.[8]

The queen of the mathematical realm was geometry and the paradigm of mathematical rigour was considered to be constituted by Euclid's *Elements.*[9] However this situation began slowly to change after the second half of the seventeenth century, due firstly to the development of algebra as a tool for solving geometrical problems in la *Géométrie* de Descartes and as a sort of generalized arithmetic[10] and secondly, to the development of infinitesimal methods. Malebranche thought,

[7] The first Latin translation of this work of Proclus is due to Francesco Barozzi and appeared in Padua in 1560.

[8] In fact Euclid was in doubt about which of the two kinds of definitions to use for the basic geometrical entities. Thus in Book I of the *Elements* he defined a point as something which has no part (definition 1) and a line as a breadthless length (definition 2). But on the other hand he defined points as the extremities of a line (definition 3), lines as the extremity of a surface (definition 3), and surface as the extremity of a solid (Book XI, definition 2) (Euclides 1956).

[9] For instance, when Torricelli in 1641 discovered that a certain solid of infinite length has a finite volume, he gave two proofs, one using his own method of indivisibles, and the other one using the method of exhaustion, a method used several times by Euclid (Mancosu and Vailati 1991).

[10] In the second edition (1689) of his *Elements of Mathematics* (*Nouveaux Élémens des Mathématiques)* father Jean Prestet wrote that algebra is the most general of all mathematical sciences and the most useful to discover mathematical truth. In his words: "Mais encore que l'Arithmétique ou la science des nombres soit une science universelle dont tant d'autres dépendent, on en explique néanmoins une autre par le moyen des lettres, qui est incomparablement plus générale et plus étendue. Cette science que je nomme Analyse, et qu'on nomme ordinairement Algèbre, sert merveilleusement à éclaircir, à étendre, et à perfectionner l'Arithmétique même et la Géométrie, et toutes les autres parties que les Mathématiques renferment. Mais ce qu'il y a de plus considérable dans cette science n'est pas son étendue et son universalité [...]. C'est la facilité que elle donne à l' esprit pour découvrir les vérités les plus cachées" (Prestet 1675, pp. 12–13).

during a part of his life, that algebra was the principal mathematical science, and Leibniz dreamed of a *Characteristica universalis*, that would be a universal mathematical science with a broader scope than those of geometry and algebra (Leibniz 1988, pp. 175–182).

In the seventeenth century, geometry became an important part of the erudite culture of the time. The proof of this can be found in the great number of books on subjects other than mathematics (Law, Metaphysics, Ethics) written in a geometrical style and form (*more geometrico*), many of which include the word "Elements" in the title. This way of thinking which considered geometry as the model of scientific knowledge, had been popularized among the educated men and, not surprisingly, came to be adopted by many within the Roman Catholic Church. In his text *Projet et Essais pour arriver à quelque certitude pour finir une bonne partie des disputes et pour avancer l'art de inventer,* written around 1677–1678 (Leibniz 1988, pp. 178–180), Leibniz gave us a list of works, that exposed *more geometrico*, on subjects that are not within the range of mathematical sciences. Leibniz said that it was the seventeenth century that made the most efforts to obtain geometrical demonstrations in all fields of knowledge.[11] Included in Leibniz' list are the following works: *Demonstratio immortalitatis animae rationalis, sive tractatus duo philosophici, in quorum priori natura et operationes corporum, in posteriori vero, natura animae rationalis explicantur* (1664), written by Kenelm Digby, a book on philosophy, which deals with the distinction between soul and body; *Euclides Metaphysicus (*1658) a book on metaphysics, written by Achates Thomas Albius; *Elementa jurisprudentiae universalis* (1660) by Samuel Pufendorf whose subject was the theory of law; *Elementa juris universe et in specie publici justinianei* (1664), also about the theory of law, written by Johan Felden; *Analysis Aristotelica ex Euclide restituta* (1658) by Erhard Weigel about the theory of science; *Renati Des Cartes Principiorum Philosophiae Pars I, et II more geometrico demonstratae (1663)* by Baruch de Spinoza about Descartes' *Principles of Philosophy.*

2 Institutional Aspects

Several scholars have investigated which were the religious institutions, whether colleges or universities, where mathematical sciences were taught. They also researched who the teachers were, what their mathematical training was and what subjects they taught. Regarding these aspects, well-known works include Krayer (1991), Romano (1999) and Paradinas (2012). There is a great deal of disagreement regarding the role played institutionally by the Order of Jesuits, for although Clavius, a member of this religious order, encouraged the teaching and learning of mathematics, his views on the importance of this science were essentially rejected in the final redaction of the *Ratio studiorum,* the document that established what

[11] Mais c ést notre siècle, qui s' est bien plus mis en frais pour obtenir des démostrations (Leibniz 1988, p. 178).

should be taught in Jesuit colleges. The first version of the *Ratio studiorum* in 1586 established a preparatory program of mathematics for the understanding of the *Second Analytics* of Aristotle, which without a certain proficiency in mathematics, were unintelligible.[12] In the second year of this program students were to use the compendium of Clavius. This cleric was a self-taught mathematician who was motivated in the study of this science by his reading of Euclid's *Elements.* However in the final version of the *Ratio studiorum,* published in 1599, mathematical teachings were reduced only to a part of the final year of the philosophy grades (dedicated to physics) where most of the teachers were supporters of the qualitative physics of Aristotle and opposed to the use of mathematics in physics. There were few Jesuits interested in teaching mathematics in schools and colleges. Frequently, in the first half of the seventeenth century, the teacher of mathematics was a novice who, after giving lessons in mathematics for 2 years, would pursue a theological formation and never have anything more to do with mathematics again, making room for another teacher to take his place. However, in the second half of the century, due to a lack of teachers, mathematics teachers began to stay in their positions for longer periods and this science would end up having some of the oldest teachers among the subjects taught at these institutions (de Dainville 1954, pp. 13–16).

In fact among the Jesuits there was no agreement on the value of the mathematical sciences neither for the education of a priest nor for the education of an *honnête homme.*[13] De Dainville cited the opinion of a Jesuit, Father Bouhier, who said that "The study of speculative sciences such as geometry, astronomy and physics is a futile amusement because these knowledges are useless and fruitless" (de Dainville 1954, p. 15). Some Jesuits had doubts about the epistemological value of mathematics. The proof of this fact was the so-called *Quaestio de Certitudine Mathematicarum*, which dealt with a quarrel about the scientific character of mathematics. Protagonists of this dispute included Pereyra, a Jesuit of Coimbra, an Aristotelian professor of natural philosophy, and, on the opposite side, Giuseppe Biancani, also a Jesuit, professor of mathematics at the University of Parma. The question discussed was if mathematical sciences are a science in the Aristotelian sense, that is, if mathematicians can prove their propositions *ex causa.* There are many good works dealing extensively with that *Quaestio*, for instance Mancosu (2008) and Sasaki (2003). Therefore we only summarize it here. Those who denied mathematics the status of science, such as Pereyra and Piccolomini, say that mathematicians do not use in their demonstrations any of the four causes recognized by Aristotle: the formal, the material, the final and the efficient causes. As an example they refer to the proof of the proposition I,32 of Euclid's *Elements* which says that the sum of the internal angles of any triangle is equal to two right angles. This proof proceeds by means of the construction of auxiliary lines and not by considering the definition of a triangle (formal cause) nor the parts of a triangle (material cause) nor the process of generating

[12] quae sine mathematicis exemplis vix possunt inteliigi (Romano 1993, p. 284).

[13] *L'honnête homme* was a model of humanity that appeared in France in the seventeenth century. *L'honnête homme* was a generalist, not a specialist. We can define it as meaning "cultivated gentleman".

this triangle (efficient cause). Another example to which they refer is that of the proofs by means of *reductio ad absurdum* which were considered proofs *a signo* and not *ex causa* because such proofs do not establish the cause that establishes a truth but only indicate that it is true. It is well known that in his *Seconds Analytics* (I,26), Aristotle considered indirect proofs or proofs by *reductio ad absurdum* of lesser value than direct proofs. The reason for this is that proofs by *reductio* are proofs *a signo*. However, the ancient Greek geometers considered indirect proofs essential both to demonstrate the existence of certain mathematical entities such as incommensurable reasons, and to determine the inexistence of others entities, such as the inexistence of a straight line lying between a circle and its tangent to a given point. Also *reductio ad absurdum* was employed by the Greek geometers in the so called method of exhaustion, used to determine areas and relationships between the areas of curvilinear figures, such as how any two circles are proportional to one another in the same ratio as the squares of their diameters (Proposition II, Book XII of the *Elements*). On the other hand, Biancani tried to show, in opposition to Pereyra and Piccolomini, that mathematical proofs agree with the Aristotelian notion of science. This quarrel demonstrates that there was no consensus, among the Jesuits, regarding the epistemological value of mathematical sciences. Several Jesuits who denied that mathematical sciences had any importance to Catholic education were proponents of Aristotle's natural philosophy. They denied that mathematics were useful to the study of nature. Other Jesuits gave more importance to a rhetorical and humanist education rather than a scientific one. But although there were disagreements among Jesuits regarding the value of mathematical sciences all of them identified themselves as Aristotelians. Their dispute was regarding whether mathematical sciences could be considered as sciences according to what Aristotle understood a science to be in his *Second Analytics*.

Let us examine summarily other religious orders. Among the Oratorians, the order of the great philosopher Malebranche, the dominant opinion regarding the value of mathematics seems to have been that of Lamy. He wrote, in the preface, to his *Élémens de Géométrie ou de la mesure de l'etendue*, that the justification for teaching this science lies not in its usefulness to different domains and activities such as astronomy, optics, physics, architecture, navigation and fortifications but in its ability to develop reasoning (Lamy 1740).[14] According to Belhoste until the half of the eighteenth century the place of mathematics in the schools run by the Oratorians was modest (Belhoste 1993). However, the important role and re-editions of their textbooks would show the contrary.[15] In any case, the teaching of mathematics gained increasing importance in the course of the eighteenth century in these schools for two reasons. The first was the gradual substitution of Cartesian physics by Newtonian physics. As it is well known, Newtonian physics required more mathematical tools than Cartesian physics. The second was the increasing use of

---

[14] Néanmoins ce n'est pas pour cela que je fonde l'estime qu'on doit faire de la Géométrie, mais sur ce qu'elle est propre pour former l'esprit et le rendre éxact, étendu et pénétrant.

[15] For example, the 1765 edition of the *Élémens des mathematiques* de Lamy is the eighth edition of that work.

mathematics in military disciplines, such as navigation, fortification and artillery. The Oratorians ran several schools preparing young men for military careers where the teaching of mathematics acquired an increasing importance (Belhoste 1993; Costabel 1964). Finally, concluding our examination, we can say that the Jansenists gave value to the teaching of mathematics in their *petites écoles*. Arnauld, Nicole and Lancelot were the authors of three textbooks, written in French and not in Latin which modernized education at secondary schools: *La Grammaire générale et raisonée*, *La Logique ou l'art de penser* and finally the *Nouveaux Elémens de Géométrie* (Schubring 2015). In a short paper Cognet affirmed that it would be too audacious to say that these books reflected any real situation of the teaching at the *pétites écoles* (Cognet 1953). However a more recent work on Jansenist education claims that Geometry was taught there in all probability from Arnauld's textbook (Sodipo 1971, p. 261).

To summarize: there was no consensus among the three main directions of the Catholic education, that of the Jesuits, that of the Oratorians and that of the Jansenists, regarding the value of mathematics for the education of a clergyman or of an *honnête homme.* Neither can we say that there was a clear institutional decision by the Catholic Church to promote knowledge of mathematics in the education of the priests of the secular clergy. However, from the time of later antiquity, the Roman Catholic Church has recognized the value of all the liberal arts for the education of a Christian man, and especially for the education of priests as we read in Saint Augustin's *de Doctrina christiana.* In that work, in book II, 40, 60, Saint Agustin wrote:

> Whatever has been rightly said by the heathen, we must appropriate to our uses. Moreover, if those who are called philosophers and especially the Platonists have said anything that is true and in harmony with our faith we are not only not to shrink form it, but to claim it four our own use from those who have unlawful possession of it [our translation].[16]

In the same book II, 39,58, Augustin said: "I think, however, there is nothing useful in other branches of learning that are found among the heathen except information about history (…), except also the sciences of reasoning and of number"[our translation].[17] On the other hand, Ignacio de Loyola, referring explicitly to mathematics, wrote "*y también las matemáticas con la moderación que conviene [deben ser tratadas] para el fin que se pretende*" (and also mathematics must be cultivated with moderation) (*apud* Paradinas 2012, p. 153). What does "moderation" mean in that context? That mathematics should not be cultivated by itself, a point of view repeated by many clergymen of Early Modernity, but as a means to reach the knowledge of Sacred Truths.

Then, how can we explain that so many priests of the Roman Church, during Early Modernity, have had such an outstanding place in the history of mathematical

[16] Ab ethnicis si quid recte dictum, in nostrum usum est convertendum. Philosophici autem qui vocantur, si qua vera et fidei nostrae accomodata dixerunt, maxime Platonici, non solum formidanda non sunt, sed ab eis etiam tamquam ab iniustis possessoribus in usum nostrum vindicanda.

[17] In ceteris autem doctrinis, quae apud Gentes inveniutur, praeter historiam rerum(…), praeter rationem disputationis et numeris, nihil utile esse arbitror.

sciences, some of them as creative mathematicians, others as promoters of this science or as philosophers of mathematics? We think that the explanation does not lie in any institutional decision made by the Roman Catholic Church and her religious orders on the value of mathematical sciences for the education of clergymen but in other four reasons: the first one being the thought among several of the members of the clergy that mathematics could be useful for accepting the mysteries of Christian faith; the second was the belief, also widespread among several members of the educated classes of Western European society of the time, that the paradigm of scientific knowledge was geometry and that this science was the only discipline that could resist and overcome sceptical challenges; the third is the thought that mathematics in some way could be useful to theology; and the fourth and last is the belief that mathematics was useful for defending and understanding certain philosophical conceptions.

3 Accepting the Mysteries of Faith

One motivation for the studying of mathematics that we can find among some of the clergymen mentioned in section one of this text is that they saw this science as a tool for defending the Christian faith. According to this view, mathematics has not intrinsic value on its own, but that familiarity with this science helped people accept the truths conveyed by the Church. This is a point of view most loudly expressed by Jansenist Antoine Arnauld. Gert Schubring has attracted the attention of researchers for the Preface to Arnauld's *Nouveaux Élémens de Géométrie* (Schubring 2015, p. 240). In that Preface, which was written by Nicole, we read the following:

> [...] between the human exercises which can most serve to incline the spirit to receive the Christian truths with less opposition and disgust, it seems that there is hardly a more appropriate than that of geometry. Because nothing is more capable of removing the soul of his attachment to the senses than an attachment to an object which has nothing pleasant to the senses. And that we can find in geometry. [our translation] (Pascal et al. 2009, pp. 96–97)[18]

However, Arnauld agreed with what was expressed by Pascal in a letter to Fermat written in August 10, 1660. Pascal argued in this text that not being a geometer was in no way a defect, however, believing geometry is a thing of great value was, as was pride in having a head full of lines, angles, circles and proportions (Pascal 1992, p. 293).[19] On the other hand, Arnauld also agreed with Pascal's point of view

[18] [...] Mais entre les excercices humains qui peuvent le plus servir [...] à disposer même l'esprit à recevoir les vérités chrétiennes avec moins d'opposition et de dégoût, il semble qu'il n'y en ait guère de plus propre que l' étude de la géométrie. Car rien n'est plus capable de détacher l'âme de cette application aux sens, qu'une autre application à un objet qui n' a rien d'agréable selon les sens; et c'est ce qui se rencontre parfaitement dans cette science.

[19] Car pour parler franchement de la Géométrie, je la trouve le plus haut exercice de l'esprit; mais en même temps je la connais pour si inutile que je fais peu de différence entre un homme qui n'est que géomètre et un habile artisan. Aussi je l'appelle le plus beau métier du monde; mais enfin ce n'est qu'un métier; et j'ai dit souvent qu' elle est bonne pour faire l'essai, mais non pas l'emploi de notre force.

according to which mathematical sciences prove that there are things whose nature cannot be understood by us, but that notwithstanding do exist. For instance, figures of infinite length and finite area such as Torricelli's hyperbolic solid (Mancosu and Vailati 1991) and incommensurable magnitudes. In this sense, geometry helps us to accept the mysteries of Christian faith which are for us incomprehensible. In their book *La Logique ou l'art de penser* the Jansenists Arnauld and Nicole wrote: "It is necessary to remark that there are things that are incomprehensible but whose existence is certain. We cannot conceive how they are and yet it is certain that they exist".[20] And they ask: "What is more incomprehensible than eternity and what is at the same time more certain?" In the same work they stated:

> The utility we can obtain from mathematical speculations is not to learn truths that on their own are rather sterile, but to know the limits of our mind in order that we confess, even reluctantly, that there are things which exist, although our spirit cannot understand them; and that it is why it is good to be weary of these subtleties, in order to subdue his presumption, and to deprive him of the boldness of ever opposing his feeble lights to the truths which the Church proposes to, on the pretext that he cannot understand them[our translation] (Arnauld and Nicole 2014, pp. 517–518).[21]

In *Nouveaux Élémens de Géométrie*, when referring to the existence of incommensurable magnitudes Arnauld affirms: "It [this existence] seems incomprehensible and indeed it is because the cause of it can only be the divisibility of matter *ad infinitum*. But it is clear that all that belongs to infinity cannot be understood by a finite mind such as ours" (Pascal; Arnauld et al. p. 315). In his work *L'esprit de la géométrie* Pascal expressed a similar point of view. According to him we do not have to deny a proposition *p* because it seems incomprehensible to us. We must, on the contrary, examine its denial *not p*, and if this is manifestly false, we must affirm *p*.

> It is a natural disease of the human being to believe that he possesses the truth directly, and from that reason comes that man is always ready to deny everything that is incomprehensible to him [...]. Therefore, whenever a proposition is inconceivable, it is necessary to suspend the judgment and not deny it, but examine the contrary proposition; and if this is manifestly false, one can then boldly affirm the first, even if it is incomprehensible[our translation] (Pascal 2008, p. 28)[22]

---

[20] Mais il faut remarquer qu' il y a des choses qui sont incompréhensibles dans leur manière, et qui sont certaines dans leur existence; on ne peut concevoir comment elles peuvent être, et il est certain néanmoins qu'elles sont (Arnauld and Nicole 2014, pp. 511).

[21] L' utilité que l'on peut tirer de ces spéculations n'est pas simplement d'acquérir ces connaissances, qui sont d'elles-mêmes assez stériles; mais c'est d'apprendre à connaître les bornes de notre esprit, et à lui faire avouer malgré qu'il en ait, qu'il y a des choses qui sont, quoiqu'il ne soit pas capable de les comprendre; et c'est pourquoi il est bon de le fatiguer à ces subtilités, afin de dompter sa présomption, et lui ôter la hardiesse d' opposer jamais ses faibles lumières aux vérités que l'Eglise lui propose, sous prétexte qu' il ne les peut pas comprendre.

[22] C'est une maladie naturelle à l'homme de croire qu'il possède la verité directement; et de là vient qu'il est toujours disposé à nier tout ce que lui est incompréhensible [...].Et c'est pourquoi toutes les fois qu'une proposition est inconcevable, il faut en suspendre le jugement et ne pas la nier à cette marque, mais en examiner le contraire; et si on le trouve manifestament faux, on peut hardiment affiremer la première, tout incopréhénsible qu'elle est.

On the other hand, Father Bernard Lamy expressed in the preface to his book *Traité de la grandeur en général*, published in 1680 that mathematics helps us conceive of spiritual realities because it detaches the mind from the things perceived by the senses and then helps us conceive of spirituals and abstracts things.[23] Moreover, Lamy said that the mathematical sciences teach us both the extension and limits of our minds. Referring to the content of his book, in the preface, Lamy said:

> This treatise reveals the extension of our mind and its limits. For there are demonstrations that clearly and convincingly prove that a finite magnitude is infinitely divisible and that there are truths, which are certain but for us incomprehensible. Consequently the truths that religion teaches us must not be suspected though they are for us incomprehensible [our translation] (Lamy 1765, xxi).[24]

4 Against Scepticism

A second motivation we have found espoused by some of the theologians of the Roman Church for studying mathematics is this: that mathematics can serve as a weapon against scepticism. It seems to be the only discipline that could overcome the sceptical doubts. For this reason Marin Mersenne dedicated three of the four parts of his book *La verité des sciences contre les Sceptiques ou Pyrrhoniens*, written in 1625, to the mathematical sciences. Referring to mathematics Mersenne said: "I want to show you that mathematics is a very certain science in which the suspension of the judgment does not find a place" (Mersenne 2003, pp. 96–97). To understand Mersenne's point of view we ought to remember that scepticism had spread in France during the sixteenth century among erudite circles, principally due to the works of Montaigne and Francisco Sanches. In the following century, François de la Mothe Le Vayer was one of the best known representative authors of French scepticism. Although the diffusion of scepticism was restricted to the domain of philosophical disputes, the fact is that it could lead to deism and ultimately to atheism. The latter opinion was shared by Mersenne, who refers to the dangers of scepticism in the preface to *La verité des sciences* saying that believing in the uncertainty of science could lead to the loss of faith (Mersenne 2003).[25] One of the mains sources

[23] Ainsi l'étude de ce traité détache davantage l'esprit des choses sensibles et donne une plus grande disponibilité pour concevoir les choses spirituelles et abstracts.

[24] Mais si ce traité fait voire l'étendue de l'esprit, il faut connoître ses bornes; car il y a des démonstrations claires et convaincentes qu'une grandeur finie est divisible à l'infini [...] ce qui démontre qu'il y a des verités qui sont également certains et incompréhensibles, et que par conséquent, les verités que la religion nous enseigne ne doivent pas être suspectées, parce qu'on ne les comprend pas entièrement.

[25] Lesquels [les libertins] n'osant faire paraître leur impiété de peur qu'il sont d'être châtiés, se efforcent de persuader aux ignorants qu'il n'y a rien de certain au monde à raison du flux et du reflux continuel de tout ce qui est ici bas; ce qu'ils tâchent de faire glisser dans l'esprit de certains jeunes hommes qu'ils connaissent être portés au libertinage et à toute sorte de volupté et de curiosité, afin que ayant fait perdre le crédit à la vérité en ce qui est des sciences et des choses naturelles qui nous servent d'échelons pour monter à Dieu, ils fassent le même en ce qui est de la religion.

of the philosophical scepticism in Early Modernity was the *Outline of Pyrrhonism*, a text written by the ancient philosopher Sextus Empiricus in Late Antiquity. Sceptics were classified into two kinds: academicians and Pyrrhonians. Montaigne in his *Apologie de Raymond Sebond* and Arnauld and Nicole in *La Logique ou l'art de penser*, wrote that the academicians believed that there are things that would be more probable than others. On the other hand, Pyrrhonians believed that all things are equally doubtful. The latter also casted doubts on men's faculties of knowledge: sensorial perception, intellectual intuition, memory and reasoning.

Apparently on account of their certainty mathematical sciences are not subject to sceptical doubts. An opinion like this had been expressed by St. Augustin in his dialogue *Contra Academicos*. "It is necessary that three times three be nine, and that nine be a square number, although all the human race is snoring" said St. Augustin refuting the sceptical doubts[26] (San Agustin 1947, p. 193). So Descartes thought, at a time of his intellectual evolution, when he wrote the *Regulae ad directionem ingenii*. In rule II Descartes said that we must deal only with those things from which we can obtain solid and indubitable knowledge. That rule limits us only to study arithmetic and geometry.[27] And that is because those sciences study such simple objects that they do not suppose anything that experience shows doubtful and consist only in deducing consequences in a rational way.[28] However, in Book I, Chap. 15 of his *Outline of Pyrrhonism*, Sextus Empiricus noted what he called the five modes leading to suspension of judgement, which are likely to appear in mathematical proofs (Sextus Empiricus 1933, pp. 94–95). The first of them is based upon discrepancy. In fact, there may be a discrepancy among mathematicians in relation to the Euclidean axioms. The discussion was about which axioms must be accepted on account of being evidently true and which require a demonstration. In Antiquity, Apollonius tried to prove the Euclidean axiom that states that two things equals to a third are equals. Doubts were also raised regarding the Euclidean postulates, for instance to the fifth postulate.[29] Sextus' second mode is based upon regress ad infinitum: the thing adduced as a proof of the matter proposed needs a further proof, and this again another, and so ad infinitum, so that the consequence is suspension of judgment, as we possess no starting point for our argument. The third mode, based upon relativity is that whereby the object has such or such an appearance in relation to the subject who judges it, but as to its real nature we suspend judgment. In fact, there were controversies among philosophers about the existence of intelligible objects, such as mathematical objects. Some say that only sensible objects are true (real), others

[26] [...] nam ter terna novem esse, et quadratum intelligibilium numerorum necesse est vel genere humano stertente sit verum.

[27] Solae supersint Arithmetica et Geometria ex scientiis iam inventis, ad quas huius regulae observatio nos reducit.

[28] Ex quibus evidenter colligitur, quare Arithmetica et Geometria caeteris disciplinis longe certiores existant, quia scilicet hae solae circa obiectum ita purum et simplex versantur, ut nihil plane supponant, quod experientia reddiderit incertum, sed totae insistunt in consequentiis rationabiliter deducendis.

[29] See (Euclid 1956, I, pp. 202–220).

that only intelligible objects are true, and others still that both intelligible and sensible objects are true. We have also seen that concerning the nature of mathematical objects there were disputes in the early modernity among those who thought mathematical objects are obtained by means of abstraction from material things and those that maintained they are innate.[30] But concerning maters of dispute which admit of no decision, says Sextus, it is impossible to make an assertion (Sextus 1933, Book I, XV). According to him, objects of thought, like the mathematical objects, are so named on account of the relation to the person thinking and, if they really possess the nature that they are said to possess, there would be no controversy regarding them.

In his *Apologie de Raymod Sebond* Montaigne casts doubts on the certainty of mathematics. Sometimes mathematical results are incompatible with sensorial experience.

> I have been told that in geometry (which claims to have reached the highest degree of certainty among scientists) there are irrefutable demonstrations which overturn truth based on experience. Jacques Peletier, for example, in my own home, told me how he had discovered two lines drawing ever closer together but which, as he could prove, would meet only in infinity. And the sole use Pyrrhonists have for their arguments and their reason is to undermine whatever experience shows to be probable (Montaigne 1987, p. 277, translated by M.A. Screech).

Besides, mathematicians, says Montaigne, proceed by assuming certain things, axioms and postulates that are undoubtedly true. But we can maintain opposite principles as well.

> Whenever a case is argued based on preliminary assumptions, in order to oppose it take the very assumption which is in dispute, reverse it and make that into your preliminary assumption. For any human assumption, any rhetorical proposition, has just as much authority as any other, unless a difference can be established by reason (Montaigne 1987, pp. 115–116, translated by M.A. Screech).

The Jansenists Arnauld and Nicole seriously considered the challenge raised by modern sceptics. The first chapter of the four parts of *La Logique* is devoted both to prove that there is science and to establish what the limits of human knowledge are. Scientific knowledge is defined by the authors in psychological and epistemological terms rather than in logical and semantical ones: science is a set of beliefs founded on strong reasons with which we cannot disagree. Sceptical arguments, say Arnauld and Nicole, are pragmatically inconsistent because they clashed with everyday life. Although the sceptics claim that we cannot know if we are asleep or awake or if there is an external world or not, in fact they behave in everyday life as if they have such knowledge. According to Arnauld and Nicole, we must distinguish between

[30] An empiricist position regarding mathematical objects was expressed in the early modernity by Gassendi (Rochot 1957, pp. 69–78).

three kinds of things. Firstly, there are things that we know with certainty. These are known through intellectual intuition (for instance, that things equals to a third one are equal between themselves) or by means of a demonstration from first principles (for instance that the three internal angles of a triangle are equal to two right angles). These things are the subject of mathematics. Secondly there are things we know with probability, as for example, the laws of nature, the existence of past facts (for instance that Julius Caesar was killed by Brutus) and the occurrences of future facts (an eclipse of the moon). Thirdly there are things that we cannot know due to the limitations of the human mind. The latter things are those referring to infinity, as for example, if an infinite number is even or odd, if the half of an infinite number is infinite, and if there are infinites greater than others (Arnauld and Nicole 2014). This classification of things is directed against the Pyrrhonian claim that all are equally doubtful and consequently we cannot know anything with certainty.

In *La verité des sciences* Mersenne deals with sceptical doubts about the existence of numbers. This work presents a dialogue between two persons, one of them named *Sceptical,* expresses the traditional Pyrrhonians views, the other, called *Theologian*, those of Mersenne. Chapter 2 of Book II of this text opens with Theologian's opinion that there is nothing in the world more certain than arithmetic because numbers measure all material and corporal things.[31] Besides, as with others immortal substances, numbers do not change, satisfying thus the Aristotelian requirement for the existence of a science, as according to this criterion science is the study of changeless things.[32] The sceptic replies that arithmetic is not certain because it is grounded on the notion of unity, which is not at all clear. In fact there is disagreement between philosophers on the nature of the unity. Some have said that the unity is a privative concept, like blindness, that denotes the negation of any manifold, and then would be nothing.[33] Then arithmetic would not be a science because only the study of real things can be a science. In Chap. 3 the sceptic tries to prove that numbers are not real. Mersenne responds that the relation between unity and the things that are one is like that of genera to species or like that of species to individuals. We will not go into the details of Mersenne's arguments nor evaluate them. We want only to show the importance given by this priest and others to mathematics within the context of the battle against modern scepticism. For Mersenne, to proving the certainty of mathematical knowledge entails a defeat of the Pyrrhonian position.

[31] On ne trouve rien de plus certain que l'Arithmétique car jamais elle ne manque en ce qu'elle entreprend; les nombres nous servent de mesure non seulement pour les choses corporelles [...] néamnoins nous les accommodons aux choses sprituelles.

[32] Les nombres n'étant plus sujets à changement que les substances immortelles.

[33] On ne sait ce que c'est l' unité, car il y en a qui dissent qu'elle n'est qu'une pure privation ou négation de multitude, tel. qu'est le point en la ligne et par conséquent l'unité n'est rien [...] Il faut que ce soit un nombre [...].

5 Usefulness to Theology

Another motivation we have found given by some clergymen of the Roman Catholic Church for studying mathematics is that it is useful to several branches of theology: biblical and moral. In his book *De doctrina christiana* Book III, Chap. 27 Saint Augustin encouraged the study of liberal arts, among them arithmetic and geometry saying that they are helpful to priests for understanding the holy scriptures.[34] St. Augustin states that "the ignorance of numbers also hinders the understanding of many things written in Scripture with a figurative and even mystical sense" (*De Doctrina*, II, 16, 25).[35] For instance, number ten means the knowledge of the Creator and the creature because three refers to the Creator (Trinity) and seven to the life and the body of the creature. In the human body are four elements and the human spirit consists of heart, mind and soul (*De doctrina*, II, 25). In Chap. 38 Augustine refers specifically to arithmetic saying that the objects with which this science deals are not creations of human minds but rather discoveries of human minds.[36]

On the use of mathematics for biblical theology Mersenne wrote in *La Verité des sciences* that mathematics is useful for understanding certain passages of the Holy Scriptures such as those referring to commutative and distributive justice, the dimensions of Noah's ark or the positions of the stars. Architecture, which is a type of applied mathematics, allows us to understand the dimensions of the temple of Solomon. Finally, geography and hydrology, which also belong to applied mathematics allow us to understand the locations of the Holy Land (Mersenne 2003, pp. 312–313).[37] Mersenne added that it is not possible to understand what the Fathers of the Church wrote, such as St. Augustin's words on the quantity of the soul in his *De quantitate animae* and about the music, in other texts, without knowledge of mathematics.[38]

Father Lamy wrote that mathematics is also useful for leading to the acceptance of Catholic moral theology. In his *Éléments des mathématiques ou traité de la grandeur en général*, Lamy wrote:

> One of the great principles of human evil is this strong inclination people have for things that are perceivable, which means that nothing pleases them other than that which flatters their senses. Thus as geometry separates from the bodies it studies all perceivable qualities

[34] Scientiarum quas homines non instituerunt, aliquae iuvant ad intelligentiam scripturarum.

[35] Numerorum etiam imperitia multa facit non intelligi, translate ac mystice posita in Scripturis.

[36] Numerorum scientia non ex hominum instituto, sed ex rerum natura est ab hominibus adiuventa.

[37] Les Mathématiques sont forts utiles pour entendre l'écriture sainte, ce qui appartient à la justice distributive et commutative. L'Arithmétique et la Géométrie enseignent clairement que l'arche de Nöe a pu contenir une couple des animaux immondes et trois couples et demie des mondes [...].L'Astronomie sert à expliquer la Prophete Job, quand il parle des Pléiades et d'Orion. L'Architecture montre l'excellence du Temple de Salomon. La Géographie et l'Hydrographie nous donnent de la lumière pour entendre les lieux de la Terre sainte.

[38] Les pères de l' Ѐglise ne peuvent être parfaitement entendus, si on ne sait la Géométrie, comme vous pouvez expérimenter par la lecture du livre que Saint Augustin a fait de la quantité de l'âme, sans mettre en ligne de conte les six livres qu'il a composé de la Musique.

and leaves them nothing which can please the concupiscence, when one can dedicate one's spirit to the study of this science, one detaches it from the senses and comes to love other pleasures than those who taste it by the means of senses [our translation].[39]

The Jesuits thought that mathematics could be useful to missionaries, especially those disciplines that were considered to belong to applied mathematics at the time, such as astronomy and geography. This knowledge was necessary for navigation and other activities related to the task of evangelization. Furthermore the intellectual elites of Eastern Asia were found to be strongly interested in all mathematical disciplines. This fact was helpful to missionaries trying to get close to these regions and cultures. However, these elites were in fact more interested in these sciences than in accepting the Christian faith, contrary to what the Jesuits had expected (Sasaki 2003).

6 Philosophical Conceptions

Among these Catholic clergymen, whether they were disseminators of mathematical knowledge by means of textbooks and commentaries on the works of the geometers of Antiquity or philosophers of mathematics, we find different attitudes both in relation to the Aristotelian philosophical tradition and to the thought of early modern thinkers like Ramus, Descartes and Pascal. Their works reflected these attitudes, whether it was in opposition to or acceptance of their ideas.

The doctrines of Protestant humanist Petrus Ramus, which spread widely throughout Europe, were highly influential during the sixteenth century. They are still recognizable in Arnauld's textbook *Nouveaux Élémens de Géométrie* whose first edition was published in 1667. Ramus wrote treatises on all the liberal arts in which he focused on the didactics of these disciplines. He wrote *Scholarum Mathematicarum* (1569) on mathematics teaching. Ramus said in his *Dialectique* of 1555, that when one is teaching a science, one must firstly explain what the object of the discipline is (de la Ramée 1996). Then one must lay down the more general propositions and move on to the more specific ones, using division of concepts (distribution) or deduction from the general principles. One must begin with the higher genera and then continue to the species. For example, if we want to explain the object of grammar we will say that grammar is the science that teaches one to speak correctly. Then we will divide it into its two species: syntax and etymology. Defining and dividing in this way we will then come to individual examples. Obviously the *Elements* of Euclid is not in accordance with Ramus' conceptions on how to teach a science. Euclid did not define the object of geometry. He stated that the discipline

[39] Un des grand principes de corruption pour les hommes est cette forte inclination qu'ils ont pour les choses sensibles, qui fait que rien ne leur plait que ce qui flatte leurs sens. Ainsi, comme la Géométrie sépare des corps qu'elle considère toutes les qualités sensibles et qu'elle ne leur laisse rien de ce qui peut plaire à la concupiscence, quand on peut forcer un esprit et obtenir qu'il s'applique à l'étudier, on le détache des sens et on lui fait connaître et aimer d'autres plaisirs que ceux qui se goutent par leur moyen.

is not concerned with space nor with continuous quantity nor any other object of which one can gain knowledge. The first four books of the *Elements* deal with plane geometry. Book V of the *Elements* addresses a more general theme: the theory of reasons and proportions whose scope is quantity in general both continuous and discrete. In Book VI, again, Euclid deals with plane geometry. In Book X he returns to a more general theme, namely, the incommensurable reasons. From Book XI onwards Euclid deals with spatial geometry. Therefore we see that it does not follow the order prescribed by Ramus. If we limit ourselves to considering only Book I of the *Elements*, we can see that in proposition I,1 Euclid solves the problem of constructing an equilateral triangle, and in proposition I,22 he shows how to construct a triangle in general. This is further evident when we look at proposition I,16 which states that the external angle of a triangle is greater than each of the opposite internal angles of the triangle, which is followed by the proposition I,32 which states that the external angle of a triangle is the sum of the two opposite internal angles. In the order prescribed by Ramus I,22 should come before I,1 and I,32 before I,16.

Arnauld accepts the methodology proposed by Ramus. He begins his *Nouveaux Élémens* with a general theory of reasons and proportions that can be applied to any type of magnitude, as long as they are homogeneous with one another.[40] Only in book V of his text does Arnauld begin the study of continuous magnitudes, which is the object of geometry. Arnauld's text breaks away from the others when he demonstrates the properties of parallel and perpendicular lines, without using angles and triangles, which differs from the Euclidean proofs. The reason for doing this is that the concepts of a triangle and angle are more complex than the concept of line, since triangles and angles are formed by lines. Therefore the order of exposition in Arnauld's text differs greatly from that of Euclid's. In the conclusion to his work this theologian wrote

> I have left other problems unsolved because it is easy to solve them using the principles I have set out. Besides having written these Elements to make a sketch of the true method that consists in treating the simplest things before the compound ones, and the general ones before the particular ones, I think I have fulfilled that purpose and have shown that the geometers erred in having neglected that natural order, imagining that they had nothing more to observe than the preceding propositions would serve to prove the following. It is made clear, by this attempt, that, by explaining the elements of geometry according to this natural order they can be incomparably more easily conceived of and remembered [our translation]. (Pascal et al. 2009, p. 761).[41]

[40] Two magnitudes are homogeneous when the lesser multiplied by a natural number can exceed the other. This concept correspond to the definitions 3 and 4 of Book V of Euclid's *Elements*.

[41] Je laisse d'autres problèmes qui sont très faciles à résoudre par les principes qui ont été établis. Outre que n' ayant entrepris ces Éléments que pour donner un essai de la vraie méthode qui doit traiter les choses simples avant les composés, et les générales avant les particulières, je pense avoir satisfait à ce dessein, et avoir montré que les géomètres ont eu tort d'avoir négligé cet ordre de la nature, en s'imaginant qu'ils n'avaient autre chose à observer, sino que les propositions précédentes servissent à la preuve des suivantes: a lieu qu'il est clair, ce me semble, par cet essai que les éléments de géométrie étant réduits selon l'ordre naturel, peuvent être aussi solidement démontrés, et sont sans comparaison plus aisés à concevoir et à retenir.

It was unanimously agreed upon among Jesuits that the study of Euclid's *Elements* was important for the comprehension of Aristotle's *Second Analytics*. In fact, during the seventeenth century the influence of Aristotelian thought was greater in Italy, Spain and Portugal, than in France, England and even Germany. After 1650, Cartesianism became the principal philosophical current in France, exerting strong influence over both the Oratorians and the Jansenists.

Those Jesuits who, like Clavius and Tacquet,[42] wrote commentaries on Euclid's *Elements*, correcting and expanding on them, accepted the Aristotelian conceptualization of geometrical definitions which we can find principally in the *Second Analytics*. However, it was not shared by all the authors we are analyzing. In this work, Aristotle distinguishes between nominal definitions and real definitions. A nominal definition gives the meaning of a term or a concept, for instance, when we define a "regular polyhedron" as the solid whose faces are regular polygons. This kind of definition says nothing about the existence of an object that would correspond with the concept or term defined. A real definition communicates the essence of a thing, for example, the definition of man as rational animal. Regarding basic geometrical entities such as a point, line and plane, Aristotle said that their definition must be given and, at the same time, their existence assumed. For other geometrical entities, like triangles and squares, their definition (nominal definition) must first be given and then their existence ought to be proven, mostly by means of a construction. This is what Euclid does in his *Elements*, because he first defines what the meaning of the term "triangle" is, and then in proposition I,1 proves the existence of the equilateral triangle and in I,22 he teaches how to construct any triangle given three straight lines. The same is done for a square: firstly the meaning of the term "square" is given and afterwards Euclid shows how to construct a square. Contrary to Euclid, the Jansenist Arnauld, due to the influence of Ramus, Descartes and Pascal, opposed giving definitions for basic geometric entities such as a point, plane or line. In his *Nouveaux Élémens de Géométrie*, he wrote:

> The ideas of a plane surface and right line are so simples that if we try to define them, it would make them confusing. What we can do is only give examples in order to link these ideas to the words of every language[43] [our translation].

Another Aristotelian principle that was widely accepted until the seventeenth century is the so-called principle of incommunicability of the genera (*Second Analytics*, I,7). According to this principle, we cannot prove a proposition relative to a kind of things (genus), said Aristotle, using considerations that belong to another kind of things. Thus we could not prove a proposition regarding geometry using concepts and tools pertaining to arithmetic or algebra because the genus of geometry is that of continuous quantities, and the genus of arithmetic is that of discrete quantities. According to Aristotle, the only exception to this principle is found in the

---

[42] The title of Tacquet's book was *Elementa Geometriae planae ac solidae quibus accedunt selecta ex Archimede Theoremata*. It was first published in 1665.

[43] Les idées de surface plane et d'une ligne droite sont si simples qu'on ne ferait qu'embrouiller ses termes en les voulant définir. On peut seulement en donner des exemples pour en fixer l'idée aux termes de chaque langue.

subordinated science, such as optics where it is possible to demonstrate a theorem of optics using geometry. It is well known that Descartes' algebraic geometry implies the negation of this principle, for the aim of the French philosopher was to use algebra to solve geometrical problems such as the famous Pappus' problem. However, the principle of incommunicability of genera was maintained, in some ways, even by those who were not confessedly Aristotelians in relation to geometrical proofs by superposition, such as the proof of proposition I,4 of the *Elements*. These proofs are based on the Euclidean axiom that states things that coincide with one another are equal to one another. Arnauld, in his *Nouveaux Élémens*, thought that the Euclidian proof of I,4 introduces considerations pertaining to physics into the field of geometry, such as movement, and therefore ought not to be included. Arnauld qualifies this Euclidean proof as crude and materialistic. He adds that this proof can only satisfy people who delight in knowing how to use their imagination instead of their intelligence, which is wrong, because their mind then becomes unable to conceive of spirituals things and considers only images as true[44] (Pascal et al. 2009, pp. 382–384).

In *La Logique ou l'art de penser*, Part IV, Arnauld and Nicole criticize the excessive use made by the geometers including Euclid, of proofs by *reductio ad absurdum*. In his *Nouveaux Éléments de Géométrie*, Arnauld tried to avoid this kind of proofs, but at the cost of greatly increasing the number of geometrical axioms. In principle, the use of infinitesimal methods such as Cavalieri's indivisibles should allow for the reduction of proofs using the method of exhaustion by substituting this kind of proof with direct proofs, but infinitesimal methods were not unanimously accepted until the end of the seventeenth century.

Descartes' influence is evident in the text book *Élémens des mathématiques* of Father Prestet, a book whose development and publication were encouraged by Malebranche. On the cover for this book we read that the work shows *une méthode court et facile pour comparer les grandeurs et pour découvrir leurs rapports par le moyen des caractères et de nombres et des lettres de l'alphabet.* The words *caractères de nombres et de lettres* refer to the use of algebra to solve geometrical and other mathematical problems, in accordance with the Cartesian approach. This is not to say that Prestet (and Malebranche) agreed completely with the Cartesian conception of algebra (Robinet 1961). In fact, according to Descartes, the role of algebra is to be a tool for geometry, thus remaining subordinate to this science. Contrarily Prestet and Malebranche considered algebra the supreme mathematical science (Robinet 1961).

The influence of the works *L'ésprit de la Géométrie* and *L'art de persuader*, written by Pascal, is visible in *La Logique ou l'art de penser* (Part IV, 10) where Arnauld and Nicole expounded the rules for the scientific method. Given that Arnauld tried to write his *Nouveaux Élémens de Géométrie* in agreement with what he himself

[44] Voilà ce qui peut satisfaire ceux qui aiment mieux se servir dans la connaissance des choses, de leur imagination que de leur intelligence: ce que je trouve fort mauvais, parce que l'esprit se rend par là incapable de bien comprendre les choses spirituelles s'accoutumant à ne recevoir pour vrai que ce qu'il peut concevoir par des fantômes et des images corporelles.

had written in his *Logic,* we find, transitively, signs of Pascal's thinking in Arnauld's work on geometry. However there are disagreements between Pascal and Arnauld relating to the value of the proofs by *reductio ad absurdum* whose use Pascal did not condemn and considered unavoidable (Pascal 2008, p. 28).[45] It is difficult to tell if there are influences from the Pascalian text *Introduction à la Géométrie,* of which only small pieces have survived, in Arnauld's *Nouveaux Élémens.*

We have discussed the philosophical influences on clergymen strongly interested in geometry. However, one of the most famous mathematicians of the seventeenth century, Father Cavalieri, of the Jesuate Order did not have any interest in philosophical questions. Regarding mathematics his attitude was that of a pure mathematician. He wrote: "I think that we should in no way spend the time that is left for us on those quarrels and disputes that are more philosophical than geometrical and which for me have always been exhausting".[46]

Regarding philosophical influences Aristotelianism was the dominant school of thought among the Jesuits, while it was Cartesianism among the Oratorians. In France these were the two most important religious orders in terms of their influence on education. At the same time, the influence of Pascal's conceptions is recognizable in Arnauld.

7 Conclusions

The majority of the members of the clergy of the Catholic Church showed, at the beginning of modernity, little interest in mathematics. The education of the Catholic clergy was based mainly on the disciplines of the *trivium*: Latin grammar, dialectics and rhetoric. Their intellectual training, as well as that of the highest social classes, was a humanistic and not a scientific one. However, a small number of clerics belonging either to the order of the Jesuits or to the Oratorians or to the Jansenist movement thought that in mathematics they had support for their religious beliefs, and they also had a means of defending the Catholic faith and of helping its propagation. Those belonging to the Order of the Jesuits remained linked to the Aristotelian philosophy. On the contrary, the Oratorians and Jansenist considered that Cartesianism was the philosophical and scientific conception that most harmonized with Christian faith.

Acknowledgements I am very grateful to Professor Gert Schubring for his criticism and comments that have improved the chapter text.

[45] See Note 20.

[46] In his enim jurgiis, et disputationibus potius philosophicis quam geometricis mihi fere semper aegrotanti, nequaquam quod superest tempus inaniter terendum esse censeo. (Exercitationes geometricae sex. Ex III).

References

Andersen, Kirsti. 1985. Cavalieri's method of indivisibles. *Archive of History of Exact Sciences* 31 (4): 291–367.

Arnauld, Antoine, and Pierre Nicole. 2014. *La Logique ou l'art de penser*. Paris: Honoré Champion.

Belhoste, Bruno. 1993. L'enseignement des mathématiques dans les collèges oratoriens au XVIII siècle. In *Le collège de Rion et l'enseignement oratorien en France au XVIII siècle*, ed. Jean Ehrard, 141–160. Paris: CNRS-éditions.

Cognet, Louis. 1953. Les Petites-Écoles de Port Royal. In *Cahiers de l'Association internationale des études françaises n°3–5*, 19–29.

Costabel, Pierre. 1964. L'Oratoire de France et ses collèges. In *Enseignement et diffusion des sciences au dix-huitième siècle*, ed. R. Taton, 67–100. Paris: Hermann.

de Dainville, François. 1954. L'enseignement des mathématiques dans les Collèges Jésuites de France du XVIe au XVIIIe siècle. *Revue d'Histoire des sciences et de leurs applications* 7 (1): 6–21.

de la Ramée, Pierre. 1996. In *Dialectique 1555*, ed. Petrus Ramus. Paris: Vrin.

Euclides. 1956. In *The thirteen books of the elements*, ed. Thomas Heath. New York: Dover.

Krayer, Albert. 1991. *Mathematik in den Studienplan der Jesuiten: Die Vorlesung vom Otto Cattenius an der Universität Mainz (1610/11)*. Stuttgart: Franz Steiner.

Lamy, Bernard. 1740. *Les Elemens de Géométrie ou de la mesure de l'etendue. Sixième édition.* Paris: Nyon Père.

———. 1765. *Elemens des mathematiques ou Traité de la grandeur en general*. Huitème edition. Paris: Despilly.

Leibniz, Gottfried Wilhelm. 1988. In *Opuscules et fragments inédits*, ed. Louis Couturat. Hildesheim: Olms.

Mancosu, Paolo. 2008. *Philosophy of Mathematics & Mathematical Practice in the Seventeenth Century*. New York: Oxford University Press.

Mancosu, Paolo, and Ezio Vailati. 1991. Torricelli's infinitely long solid and its philosophical reception in the seventeenth century. *Isis* 82: 50–70.

Mersenne, Marin. 2003. *La verité des sciences contre les Sceptiques ou Pyrrhoniens*. Paris: Honoré Champion.

Montaigne, Michel de. 1987. *An Apology of Raymond Sebond*. Trans. M. Screeech. London: Penguin Books.

Paradinas, Fuentes Jesús Luis. 2012. Las matemáticas en la *Ratio studiorum* de los jesuitas. *Revista Llull* 35 (75): 129–162.

Pascal, Blaise.1992. *Oeuvres complètes*, tome 3, ed.Jean Mesnard. Paris: Desclée de Brouwer.

———. 2008. *L'esprit de la Géométrie. De l' art de persuader*. Paris: Bordas.

Pascal, Blaise, Antoine Arnauld, and François de Nonancourt. 2009. *Géométries de Port Royal*. Paris: Honoré Champion.

Prestet, Jean. 1675. *Elémens de mathématiques ou principes generaux de toutes les sciences qui ont les grandeurs pour objet*. Paris: André Praland.

Robinet, André. 1960a. Jean Prestet ou la bonne foi cartésienne. *Revue d'histoire des sciences et de leurs applications* 13 (2): 95–104.

———. 1960b. Le groupe malebranchiste introducteur du Calcul infinitésimal en France. *Revue d'histoire des sciences et de leurs applications* 13 (4): 287–308.

———. 1961. La philosophie malebranchiste des mathématiques. *Revue d'histoire des sciences et de leurs applications* 14 (3–4): 205–254.

Rochot, Bernard. 1957. Gassendi et les mathématiques. *Revue d'histoire des sciences et de leurs applictions* 10 (1): 69–78.

Romano, Antonella. 1993. À propos des mathématiques jésuites: Notes et réflexions sur l'ouvrage d'Albert Krayer, *Mathematik im Studienplan der Jesuiten. Revue d' Histoire des Sciences* 46: 281–292.

———. 1999. *La Contre-Réforme Mathématique. Constitution et Diffusion d'une Culture Mathématique à la Reanissance*. Rome: École Française de Rome.

San Agustin. 1947. *Contra los académicos*. Bilingual edition. Trans. V. Capanaga. Madrid: Biblioteca de autores cristianos.

Sasaki, Chikara. 2003. *Descartes's mathematical thought*. New York: Springer.

Schubring, Gert. 2015. From the few to the many: On the emergence of mathematics for all. *Recherches en Didactique des Mathématiques* 35 (2): 221–260.

Sextus, Empiricus. 1933. *Outlines of Pyrrhonism*. Trans. R.G. Bury. Cambridge, MA: Harvard University Press.

Sodipo, Abrimala. 1971. *Jansenist Education in the petites écoles of Port Royal* (1637–1660) .Ph D Thesis. University of Sussex.

Taussig, Sylvie. 2018. Marin Mersenne, moine, mathématique et philosophe. http://journais.openedition.org/bibnum.

Chapter 4
"Je n'ai point ambitionnée d'être neuf": Modern Geometry in Early Nineteenth-Century French Textbooks

Jemma Lorenat

Abstract This article aims to show how early nineteenth-century French geometry textbooks incorporated concepts from modern geometry. As will be shown, textbook authors in this time period rarely incorporated new developments from research mathematics into their teaching material. Modern geometry could only enter textbooks when authors had opportunities to learn new research and were willing to challenge the increasingly prescribed state geometry curriculum. Finally, and most importantly, the types of modern geometry that entered textbooks had to have perceived value for a student audience. A systematic study will illustrate how pedagogical values shaped the presentation and integration of modern geometry in ways that persisted through twentieth-century iterations.

Keywords Modern geometry · Practical geometry · Projective geometry · Nineteenth-century French mathematics

1 Introduction

The early nineteenth century was a fertile time for geometry research in France. New institutions like the École Polytechnique and the *Annales des mathématiques pures et appliquées* encouraged and published findings. In the mid-1820s, these advances were chronicled by Auguste Cournot in the *Bulletin des sciences mathématiques, astronomiques, physiques et chimiques*.

> It is not off-topic to call the attention of our readers to the progress that geometric studies have made in recent times. Long neglected for mathematical research of another kind, pure geometry, this elder sister of all the sciences, is newly in favor; new and keen studies bear fruit. Whereas, in what one calls analysis, behind an apparent richness of procedures and methods, good minds have discovered real poverty (so that often the importance of applications can only compensate for the dryness of the work), the elegant and varied results, which the science of extension enriches each day, show what an inexhaustible mine of

J. Lorenat (✉)
Pitzer College, Claremont, CA, USA
e-mail: Jemma_Lorenat@pitzer.edu

G. Schubring (ed.), *Interfaces between Mathematical Practices and Mathematical Education*, International Studies in the History of Mathematics and its Teaching,
https://doi.org/10.1007/978-3-030-01617-3_4

> research this simple notion opens for the human mind. Distinguished geometers, among whom one must cite MM. Gergonne, Poncelet, Steiner and several others, have understood that, in order to advance science, one must leave at once both the methods of Greek geometers and this geometry, called analytic, which has only truly embraced a very limited part of the theory of extension. (Cournot 1827, p. 298)[1]

Many of these cited geometers who advanced pure geometry also expanded analytic geometry, often translating the objects of pure geometry into coordinate equations.[2] At the same time, new journals of mathematics emerged enabling authors to more quickly put their results into print and before multiple audiences. While books continued to appear, authors increasingly relied on the faster and more convenient article format to publicize and propagate their new findings and methods.

The beginning of modern geometry, as Poncelet called it (Poncelet 1817)—or projective geometry, as it would come to be called[3]—is well documented in the historical literature. Beginning in 1837 with Michel Chasles' *Aperçu Historique*, through mid-century necrologies and biographies, to the geometry articles of Felix Klein's *Encyklopädie der mathematischen Wissenschaften*, nineteenth-century geometers were eager to trace the historical development of their research (Chasles 1837; Fano 1907; Schoenflies 1909; Loria 1887).

In the mid-twentieth century, René Taton situated the emergence of "modern geometry" in the early nineteenth century. Taton recognized that certain defining aspects of modern geometry (under different terminology) dated back to ancient times, including the study of conjugate diameters with Apollonius and anharmonic ratios with Pappus. Further, modern geometry owed its origins to the foundational work of Girard Desargues and Gaspard Monge, particularly with infinite and imaginary points. Yet, Taton considered these geometers, along with Lazare Carnot, as constituting the prehistory of modern geometry:

> We will here limit this study of the prehistory of modern geometry because the study of the work of the disciples of Monge and Carnot already belongs to the history of this branch of

[1] Il n'est pas hors de propos d'appeler l'attention de nos lecteurs sur les progrès qu'ont faits dans ces derniers temps les études géométriques. Long-temps délaissée pour des recherches mathématiques d'un autre ordre, la géométrie pure, cette soeur aînée de toutes les sciences, reprend une faveur nouvelle; des aperçus nouveaux et piquants viennent la féconder. Tandis que, dans ce qu'on appelle analyse, derrière une apparente richesse de procédés et de méthodes, de bons esprits ont découvert une pauvreté réelle (en sorte que souvent l'importance des applications peut seule compenser l'aridité du travail), les résultats élégants et variés, dont s'enrichit chaque jour la science de l'étendue montrent assez quelle mine inépuisable de recherches cette notion si simple ouvre à l'esprit humain. Des géomètres distingués, parmi lesquels il faudrait citer MM. Gergonne, Poncelet, Steiner et plusieurs autres, ont compris que, pour faire avancer la science, il fallait sortir à la fois et des méthodes des géomètres grecs, et de cette géométrie, dite analytique, qui n'embrasse vraiment qu'un côté fort restreint de la théorie de l'étendue.

[2] The analytic geometers best known from this time period include Charles Dupin, Joseph-Diez Gergonne, August Möbius, and Julius Plücker. Carl Boyer's chapter on the "Golden Age of Geometry" in Boyer (1956) provides a helpful overview of this period. For more detailed information, see Clebsch (1872), Otero (1997), and Gérini (2010).

[3] "Projective geometry" was first coined by Olry Terquem in 1859 as one of eight geometries that exist today "distinguées les unes des autres par des différences logiques" (Terquem 1859).

geometry whose developments were so rapid and fruitful over the course of the 19th century. (Taton 1949, p. 212)[4]

Subsequent historians have continued to study the "disciples of Monge and Carnot" and developments in the study of projective properties (Nabonnand 2011, 2015; Lombard 2011; Friedelmeyer 2011), ideal and imaginary objects (Rowe 1997; Nabonnand 2016), and the principle of duality (Chemla 1989; Chemla and Pahaut 1988).

While these studies focus on the history of geometrical research, by the mid-1860s entire textbooks on modern geometry were used for teaching in higher education courses throughout Western Europe and the United States into the twentieth century (Housel 1865; Reye 1866; Cremona 1873; Mulcahy 1862). Unlike descriptive geometry, which was first disseminated in the classroom, the process of establishing the so-called modern geometry as a standard course of study was gradual, multifaceted, and led to numerous iterations (projective geometry, geometry of position, etc.).[5] Nevertheless, both descriptive and modern geometries remained closely linked with similar modes of evolution. As Evelyn Barbin has documented in her study of how descriptive geometry changed over the nineteenth century, "journals are a good vehicle to move ideas between teachers of *Classes Préparatoires*, and to propagate new methods among secondary schools teachers" (Barbin 2015, p. 59). Similarly, we will see how concepts introduced in journals motivated changes in pedagogical content with respect to teaching modern geometry.

However, as Bruno Belhoste warns in "Pour une réévaluation du rôle de l'enseignement dans l'histoire des mathématiques," teaching modern geometry was not simply a process of reproduction from the research context:

> This is because most still consider the communication, transmission, and popularization of mathematical knowledge as secondary and peripheral activities. Under this indifference hides in fact the false idea that mathematical production can be separated a priori by the historian from the conditions of its reproduction. (Belhoste 1998, p. 289)[6]

---

[4] Nous limiterons ici cette étude de la préhistoire de la géométrie moderne, car l'étude de l'oeuvre des disciples de Monge et Carnot appartient déjà à l'histoire de cette branche de la géométrie dont les développe-ments furent, au cours du XIXe siècle si rapides et si fructueux.

[5] The difficulty of determining what this subject should be called is exemplified in Luigi Cremona's introduction:

Various names have been given to this subject of which we are about to develop the fundamental principles. I prefer not to adopt that of Higher Geometry (Géométrie supérieure, höhere Geometrie), because that to which the title "higher" at one time seemed appropriate, may today have become very elementary; nor that of Modern Geometry (neuere Geometrie), which in like manner expresses a merely relative idea; and is moreover open to the objection that although the methods may be regarded as modern, yet the matter is to a great extent old. Nor does the title Geometry of position (Geometrie der Lage) as used by STAUDT seem to me a suitable one, since it excludes the consideration of the metrical properties of figures. I have chosen the name of Projective Geometry, as expressing the true nature of the methods, which are based essentially on central projection or perspective. And one reason which has determined this choice is that the great PONCELET, the chief creator of the modern methods, gave to his immortal book the title of *Traité des propriétés projectives des figures* (1822). (Cremona 1885)

[6] C'est que la plupart considèrent encore la communication, la transmission, et la vulgarisation du

Indeed, the content and practices of teaching modern geometry by the late nineteenth century reflected an evolution and active restructuring of the subject that only resembled cited foundational texts, like Poncelet's *Traité des propriétés projectives*. Rather than a one-way transmission, teaching practices were shaped by decades of feedback among overlapping audiences and contributors. Looking back to the beginning of the century, early research in modern geometry was never far from teaching. Journals like the *Correspondance sur l'École Polytechnique*, the *Annales*, and the *Journal der reine und angewandte Mathematik* included posed problems to encourage students to apply new methods and engage in accessible research.[7] Some of the most famous modern geometers—Poncelet, Gergonne, Plücker, Steiner — often cited their teaching experiences in their publications.

In this chapter, I will focus on the adaptation of modern geometry within French textbooks during the first third of the nineteenth century. This study comprises a small fraction of the process of developing autonomous modern geometry courses, which spanned diverse temporal and local variations. Even so, documenting modern geometry in French textbooks will illustrate how pedagogical values shaped the presentation and integration of modern geometry in ways that persisted through twentieth-century iterations.

2 Historiography of Mathematics Textbooks

The historical study of textbooks in nineteenth-century France has expanded greatly over the past three decades. In Jean Dhombres' 1985 statistical study of "French mathematical textbooks from Bézout to Cauchy," he shows that, as compared with other contemporary disciplines, mathematical textbook writing constituted a significantly larger proportion of mathematical writing than research articles (Dhombres 1985). Dhombres explains the proliferation of textbooks with respect to three of the four roles of mathematics in the period from 1775 to 1825:

> First, mathematics was a favored field of education since the Revolution, appreciated both by students and by educators. We have already given statistical figures for students. The aims of teachers might have been different, but to discuss it requires the analysis of the content of the textbooks, which we postpone to another paper. Second, mathematics was a means of selecting candidates to higher positions through an elitist process, which nonetheless satisfied the egalitarian ethos of the Revolution. This process tended to be more and more organized in various fields with the model being the *École Polytechnique*. This elitist process, via a selection based on mathematics, has already begun with the military schools organized after 1770, but it obtained its peak when only an exam on mathematics was required to enter the *École Polytechnique*. Third, mathematics appeared as the necessary

savoir mathématique comme des activités secondaires et périphériques. Sous cette indifférence se cache en fait l'idée fausse que la production mathématique peut être séparée a priori par l'historien des conditions de sa reproduction.

[7] Through the early nineteenth century, posed problems attracted a diverse range of respondents. See Despeaux (2008), Gérini and Verdier (2007), Rollet and Nabonnand (2013), Delcourt (2011).

> language for developing all other sciences (Condillac used the expression "la langue des calculs", which became the title of one of his posthumous books published at the end of the Revolution). (Dhombres 1985, p. 116)

Indeed, all three roles (the fourth is mathematics for its own sake) motivated the inclusion of modern geometry within the textbook literature.

Further, Dhombres points out that "mathematical books reached a far larger audience than mathematicians" (136). The fact that textbooks are emblematic of the wider public facing side of mathematics presents another tool toward understanding which parts of modern geometry were assimilated into textbook literature. Textbook authors could determine which parts of mathematical research seemed most valuable and appropriate for more general consumption. For instance, geometry textbooks reveal an absence of imaginary or ideal objects.

While Dhombres overviews the diversity of textbooks in a given location and time period, Gert Schubring focuses his 1987 study on "Lacroix as Textbook Author" in order to outline a methodology of textbook analysis. He proposes a three-dimensional historical scheme beginning with comparing a single textbook across multiple editions, then examining corresponding changes in contemporary textbooks in light of "changes in the syllabus, administerial decrees, didactical debates, evolution of mathematics, changes in epistemology, etc." (Schubring 1987, p. 45).

The institutional factors behind textbook production are further examined by Belhoste and Renaud d'Enfert, respectively. In *Les Sciences dans l'enseignement secondaire français: textes officiels*, Belhoste explores how government policies shaped the teaching of science between 1789 and 1914. During the first third of the nineteenth century, mathematics education remained fairly static as a result of officially sanctioned texts and institutional entrance exams. The standards for mathematics education were set in Paris, and up until 1840 many decisions with respect to instruction were determined by the mathematician Siméon-Denis Poisson, who served as a member of the *Conseil Royal de l'Instruction Publique* from 1820 to 1840.

> True «patron» of mathematics in France, he is at once the exit examiner for the *École Polytechnique*, which allows him to keep an eye on the preparatory course, and the president of the jury of the science agrégation, which assures him the control of recruiting mathematics and physics teachers. (Belhoste 1995, p. 30)[8]

Thus, it is no surprise that most textbooks were written by Parisian authors, often associated with the École Polytechnique, and for purposes of exam preparation (see the Appendix for publication data).

By contrast, d'Enfert portrays regional variation in his study of mathematics education for workers. Although the movement to provide regular evening courses

[8] Véritable «patron» des mathématiques en France, il est. à la fois examinateur de sortie à l'École Polytechnique, ce qui lui permet d'avoir un oeil sur la filière préparatoire, et président du jury d'agrégation des sciences, ce qui lui assure le contrôle du recrutement des professeurs de mathématiques et de sciences physiques.

on geometry and mechanics for French workers initiated with Charles Dupin in Paris, by 1830 records show enrollment of 4000 to 5000 students in 109 towns:

> A variety of local situations respond to this general movement. This variety also affects the nature of teaching dispensed by the teachers themselves. All the courses instituted in the second half of the decade 1820 were not exactly modeled on Dupin. (D'Enfert and Fonteneau 2011, p. 89)[9]

While these instructors had more liberty to personalize their courses than their counterparts in formal education, d'Enfert observes that including "more theoretical new mathematical knowledge" in this practical context was perceived as controversial (99). Textbooks in this genre could thus only incorporate modern geometry insofar as it could be useful to the intended audience.

Most recently, Guillaume Moussard has investigated the circulation of problems and methods within elementary and analytic geometry textbooks in France between 1794 and 1891 (Moussard 2015). Of particular interest here is his chapter "L'essor de la géométrie rationnelle: nouvelles notions et méthodes" on how new geometrical notions and methods informed two textbooks on teaching geometry to workers during the 1830s. Moussard concludes that during this period modern geometry (what he calls "géométrie rationelle") did not enter secondary teaching in the strictly regulated lycées or colleges.

> Finally, we will research the presence of this rational geometry in secondary teaching. We will see that we find it less in the classical teaching of Lycées and Collèges than in industrial teaching, where the geometry teaching texts of Claude Lucien Bergery in 1826 and Étienne Bobillier in 1832 incorporated numerous elements. (Moussard 2015, p. 68)[10]

This article will be similar in that it also examines the presence of new "notions" in geometry textbooks. However, while Moussard compares methods for teaching geometry from the late eighteenth to early twentieth centuries, here modern geometry will be compared synchronously across a range of textbooks. In complement to Moussard's findings, I will examine multiple motivations for how and why different early nineteenth century authors introduced, situated, and changed certain objects from modern geometry.

All mathematics education remained fairly conservative due to strict centralized content regulations and unchanging standards of admission through the first half of the nineteenth century. In a summary of mathematics education in France from 1800 to 1980, Hélène Gispert discusses the initially bifurcated French education system, where secondary schools taught theoretical mathematics and primary schools taught practical mathematics. In both these situations, mathematics above

[9] À ce mouvement d'ensemble répond la variété des situations locales. Cette variété concerne aussi bien la nature de l'enseignement dispensé que les professeurs eux-mêmes. Tous les cours institués dans la seconde moitié de la décennie 1820 ne sont pas exactement calqués sur le modèle de Dupin.

[10] Ensuite, nous recherchons la présence de cette géométrie rationnelle dans l'enseignement secondaire. Nous verrons que nous la trouvons moins dans l'enseignement classique des Lycées et Collèges que dans l'enseignement industriel, où les ouvrages d'enseignement de la géométrie de Claude Lucien Bergery en 1826 et d'Étienne Bobillier en 1832 en intègrent de nombreux éléments.

the elementary level was considered accessory to other subjects. Mathematics only "occupied an important place in the specialized courses that were offered, often in private institutions [...] and that prepared for the écoles spéciales of the government, of which the École Polytechnique held the highest rank" (Gispert 2014, p. 230). As Caroline Ehrhardt observes in her study of algebra education,

> In spite of successive reforms about the general scientific training in high schools between 1808 and 1830, the program of the mathematical courses for students who wanted to make a scientific career remained mostly unchanged from the first years of the century to the 1830s. (Ehrhardt 2010, p. 93)

Studies of early nineteenth-century textbooks show that the most prolific geometry textbooks were those prescribed by the government and authored by Silvestre-François Lacroix and Adrien-Marie Legendre. Between 1799 and 1832, Lacroix's *Élémens de géométrie* and Legendre's *Éléments de géométrie* each ran fourteen editions with little change in content.[11] Consequently, many mathematics teachers of the 1830s essentially taught from the same textbooks that they had learned from as students.

Charles Dupin claimed "Les progrès de la science ne sont vraiment fructueux, que quand ils amènent aussi le progrès des Traités élémentaires" (Dupin 1813). This sentiment was far from universal. In fact, as will be shown, textbook authors during the early nineteenth century rarely incorporated new developments from research mathematics into their teaching material. Modern geometry could only enter textbooks when authors had opportunities to learn of new research and were willing to challenge the increasingly state prescribed geometry curriculum. Finally, and most importantly, the types of modern geometry that entered textbooks had to have perceived value for a student audience.

3 Finding Textbooks

To identify the presence of modern geometry in textbooks, this article will focus on the presence of new research objects in geometry. Admittedly, this is a rather conservative marker and may miss certain textbooks with subtler forms of modern geometry, such as the theory of transversals following Carnot. However, as will be shown, almost every textbook that emphasized new content in the introduction also included some of the new objects from research publications in the body of the text.[12] Further, this criterion coincides with the observed pattern in contemporary research articles, in which geometers praised and adopted new vocabulary in advance of new methods or theories.

---

[11] The perceived values of these two textbooks and the relationship between their authors are described in Schubring (1987).

[12] The exception here is Charles Dupin, who introduced his own new objects within his textbooks that later became part of differential geometry.

> Analysts perceiving that certain quite complicated functions are reproduced frequently in their calculations, have called them exponentials, logarithms, sines, tangents, factorial derivatives, etc.; they have created abbreviated signs to designate them, and their formulas have acquired greater clarity and conciseness. And thus for certain points, certain lines and certain circles whose consideration is frequently represented in geometric speculations, it is natural to do the same with respect to them, and to call them, following their properties, similitude centers, radical centers, polars, similitude axes, radical axes, circles of common power, etc. This attention must inevitably introduce analogous simplifications in the statement of theorems and in the solution of problems, which belong to the science of magnitude. (Anonymous 1827a, p. 279)[13]

This quote from an anonymous *Bulletin* review of Steiner, provides a list of new objects that emerged in the *Journal de l'École Polytechnique* (radical axes (Gaultier 1813)), *Correspondance sur l'École Polytechnique* (similitude centers (Hachette and Monge 1813)), *Annales des mathématiques pures et appliquées* (polars (Servois 1810)), and *Journal für die reine und angewandte Mathematik* (circles of common power (Steiner 1826)). These objects propagated through research articles, often independently from the methodological context in which they first emerged.

Significantly, most of these objects persisted into the textbooks of the twentieth century. Thus, though not capturing all of the ways in which modern pure geometry might transition from research to teaching, the paths of new objects tell significant and enduring accounts in the story. The use of poles, polars, similitude, and radicals signaled a foray into the modern geometry of the early nineteenth century that would later be characterized as projective geometry.[14]

To obtain an appropriate corpus of contemporary geometry textbooks, I first queried the *Bibliothèque nationale de France* library catalog for all texts that included the keyword "Géométrie" and had been published between 1800 and 1833 (www.bnf.fr). This search returned 113 available texts, some of which were multiple editions of the same title.[15]

Certainly, this form of search did not gather every single book on geometry published in French between 1800 and 1833.[16] Nevertheless, this search appears to be

---

[13] Les analystes s'étant aperçu que certaines fonctions assez compliquées se reproduisaient fréquemment dans leurs calculs, les ont appelées exponentiels, logarithmes, sinus, tangentes, dérivées factorielles, etc.; ils ont créé des signes abréviatifs pour les désigner, et leurs formules en ont acquis beaucoup de clarté et de concision. Puis donc qu'il est. certains points, certaines droites et certains cercles dont la considération se représente fréquemment dans les spéculations de la géométrie, il est. naturel d'en user de même à leur égard, et de les appeler, suivant leurs propriétés, centers de similitude, centers radicaux, polaires, axes de similitude, axes radicaux, cercles de commune puissance, etc. Cette attention doit introduire inévitablement des simplifications analogues dans l'énoncé des théorèmes et dans la solution des problèmes qui appartiennent à la science de l'étendue.

[14] For instance, in David Eugene Smith's very brief *History of Modern Mathematics* he points to "the theory of the radical axis" as one of several contributions that affected elementary geometry during the nineteenth century (Smith 1906).

[15] Several texts were listed in the BnF catalog, but reported "hors usage," and thus could not be accessed.

[16] For example, Poncelet's 1822 *Traité des propriétés projectives* was not found in this search because this first edition did not receive any classification and the word "géométrie" is cut-off from

representative, which I confirmed by conducting the same search through the Library of Congress online catalog (http://catalog.loc.gov/). The Library of Congress search added one additional text, the 1812 edition of Étienne Bézout's *Cours de mathématiques* originally published in 1772 (Bézout and Reynaud 1812). The same search through the Catalogue collectif de France (http://ccfr.bnf.fr/portailccfr/jsp/index.jsp), not including the Bibliothèque nationale, returned 18 new texts, of which I was able to consult 12.

In this chapter, textbooks will be defined as books that explicitly advertised to an audience of teachers or students through the title, subtitle, dedication, preface, or introduction. Acknowledging that other books might still have been used in classrooms or for self-study, this criterion applied to 79 of the 113 texts. Thus, a direct reference to the intended audience was a fairly common practice. For instance, the 1803 edition of Lacroix's analytic geometry textbooks contained a page listing all his textbooks included in the "Cours de Mathématiques pures, à l'usage de l'École centrale des Quatre-Nations" (Lacroix 1803b). In contrast to this formality, Alexandre Vincent dedicated his 1826 *Cours de géométrie élémentaire* to "students."

> This work belongs to you in more than one way: it is for you, it is with you that I wrote it: receive its dedication. May it nourish in you, as you recall the hours of our meetings, that love of study that will soon set you as well (I hope) to pay the tribute you owe to public utility. (Vincent 1826, p. i)[17]

In general, textbooks so-defined were written for teachers to use with their students, or, less frequently, for immediate student consumption.

The cost of production may help to explain why so few books appeared that weren't textbooks, and why only textbooks were reprinted in quick succession. Many of the well-known and widely republished names in turn of the century geometry—Monge, Lacroix, Legendre—wrote books almost exclusively for a student audience. Textbooks catered to an existent market, while research books were expensive and risked not being sold.[18]

Most textbook titles indicate their subject as elementary geometry, elementary analytic geometry, descriptive geometry, or practical geometry. This is in marked contrast to research articles. For instance, there were no courses corresponding to the popular *Annales* subject headings: *Géométrie de la règle*, *Géométrie de situation*, *Géométrie transcendante*, *Géométrie pure*, *or Géométrie des courbes et surfaces*.

---

the full title within the library catalog, it reads "Traité des propriétés projectives des figures..." The 1865 editions were classified as Géométrie descriptive and the full title is printed, thus these do show up if there is no date restriction.

[17] Cet ouvrage vous appartient à plus d'un titre: c'est pour vous, c'est avec vous que je l'ai composé: recevez-en la dédicace. Puisse-t-il, en vous rappelant les heures de nos entretiens, alimenter en vous cet amour de l'étude qui vous mettra bientôt à même (je l'espère) de payer le tribut que vous devez à l'utilité publique.

[18] The cost of production has been studied by Norbert Verdier in his thesis on Liouville's Journal (Verdier (2009). Jean and Nicole Dhombres addressed these issues from the perspective of books, and particularly textbooks in Dhombres (1985) and Dhombres and Dhombres (1989).

In this corpus, only the textbooks of Olry Terquem, who emphasized his different approach, proposed introducing geometry alongside algebra (Terquem 1829). Otherwise, analytic geometry was the next most advanced geometry, to be learned by those who mastered both elementary geometry and algebra, and continued to pursue mathematics. Descriptive geometry appeared after elementary geometry, either before or after analytic geometry. Finally, practical geometry was for a different group of students, often industrial workers in public courses, and might serve as their only mathematics training beyond basic arithmetic.

For each of the 79 texts, I consulted the title page, table of contents, any prefatory remarks, and the sheets of figures (nearly always located at the very end of the volume).[19] When the table of contents or introduction included any of the new objects cited above, referenced recently published articles, or broadly mentioned new geometric content then I included the text as part of my corpus of textbooks containing modern geometry. This turned out to be a very small corpus of only seven titles, several in multiple editions.

To understand why modern geometry entered textbooks, I will first consider how textbooks justified their existence and attracted readers through claims of novelty. For the majority of textbooks, novelty was framed in terms of pedagogical values. By contrast, in the seven textbooks that did contain modern geometry, authors also emphasized the novelty of the content. I will then take a closer look as to how authors developed specific aspects of modern geometry within a teaching context, simultaneously extending the tools for learning geometry while remaining within the bounds of constructive practices.

4 Claims for Novelty By Textbook Authors

4.1 The Majority View

Most claims for novelty in textbooks concerned best teaching practices. Authors debated whether theorems should appear before or after their proofs, whether problems should be embedded in the text or collected in an appendix (Develey 1812; Legendre 1800; Vincent 1826); the appropriate use of proof by contradiction (Lacroix 1803a; Schwab 1813; Olivier 1835), and how much rigour could be obtained without sacrificing the more important quality of simplicity (Lacroix 1799; Vincent 1826; Develey 1812; Clairaut 1830; Mutel 1831; Terquem 1829, etc.). Distinct forms of teaching could be subtle but were still advertised, such as the decision by Louis-Etienne Develey, Auguste Mutel, and Vincent to state propositions without reference to the lettered figure, in order that the wording might more easily be committed to memory, which all three highlighted as important decisions in their

[19] Unfortunately, when consulting scanned texts, the figure pages were often poorly copied. While disappointing, this feature was in general not a detriment toward understanding the book's content nor the author's textual use of figures.

introductions. As a further example, the Abbé de la Caille allowed his text to be more or less advanced through restricting "less useful or less easy" material to small font that the reader could include or ignore depending on preference (de LaCaille and Labey 1811, 1741, p. iv).

Within their introductions, authors both acknowledged and criticized the work of contemporary textbook writers, such as when Develey described the ongoing dialog on the best form of presenting the elements:

> A lot has been written on the best form to give to the Elements of Geometry; I do not wish to repeat what others have said and very well for I could not do it. But with these excellent directions, do we achieve perfect Elements? I do not think so; and I am far from believing that mine are thus. Several authors have taken great steps toward this perfection as we see everything in perspective; I have also attempted some efforts; perhaps one day someone luckier, but above all abler than I, will achieve the desired goal. (Develey 1812, p. v)[20]

Authors often described their work as supplementing rather than replacing previous treatments. Antoine Charles Poullet-Delisle assured the reader that his publication should not be perceived as a criticism of contemporaries, and only intended to be useful. He professed: "I have no ambition to be new: in a work of this kind that would be undoubtably a ridiculous pretension" (Poullet-Delisle 1809, p. v).[21]

When evaluating who had succeeded in writing geometry, Legendre was portrayed as the standard. Legendre himself began each new edition by thanking the various geometers who had recently offered new and relevant material including over the years Lhuilier, Cauchy, and Querret (Legendre 1800, 1812, 1832). Although feedback from other mathematicians could be useful, the ultimate test of a text's success, as Biot observed, was by experiment, "test it on the minds of the students, and verify by this proof the goodness of the chosen methods" (Biot 1810, p. vi).

The expression "modern" possessed a more traditional connotation in most textbooks, particularly those with editions dating back to the eighteenth century. As elementary geometry was considered the "method of the ancients," so analytic geometry was considered "modern." Bossut, whose text originally appeared in 1772, described analytic geometry as producing a "revolution" in "the empire of mathematics" (Bossut 1800, p. xii). Late eighteenth and early nineteenth-century geometers credited the origin of this modern geometry to Viète and Descartes, admired the work of Newton, and were inspired by both the form and content of Euler's trigonometric and analytic texts. For instance, citing Viète, Descartes, Newton, Euler, and Cramer, Lacroix provided a brief history of analytic geometry, which he prefaced in praising the "moderns."

[20] On a beaucoup écrit sur la meilleure forme à donner aux Élémens de Géométrie; je ne voudrais pas répéter ce que d'autres ont dit, et bien mieux que je ne pourrais le faire. Mais avec ces excellentes directions, sommes-nous parvenus à avoir des Élémens parfaits? Je ne le pense pas; et je suis bien loin de croire que les miens le soient. Quelques auteurs ont fait de grands pas vers cette perfection que nous voyons tous en perspective; j'ai voulu hasarder aussi quelques efforts; peut-être un jour quelqu'un plus heureux, mais surtout plus habile que moi, atteindra-t-il le but désiré.

[21] Je n'ai point ambitionnée d'être neuf: dans un ouvrage de cette espèce, ce serait sans doute une prétention ridicule.

> Then came the application of algebra to geometry; this branch, due entirely to the moderns, and whose discovery soon gave them a huge advantage over the ancients, had to change form in measure as it was extended and perfected. (Lacroix 1803b, p. vi)[22]

Yet citations back to the seventeenth century suggest that claims to modernity in analytic geometry did not necessarily imply recent development nor attention to new research. Algebraic solutions that indicated imaginary, infinite, and to some extent negative points or curves were usually dismissed as impossible or absurd.[23] Solutions that could not be represented on paper were non-existent. In fact, as will be shown, new research was just as infrequently adapted to analytic geometry as to any other geometry textbook.

4.2 Textbooks with Modern Geometry

The presence of modern geometry from contemporary research coincided with markedly different claims for novelty among textbook writers. A chronological introduction to the authors, titles, and circumstances of publication will provide a background against which such claims can be better evaluated.

Dupin

Charles Dupin (1784–1873) is both the epitome and the exception among the other authors in this study. His commitment to developing pure and analytic methods within research and teaching provided him with a remarkable professional status among his contemporaries exhibited by citations and dedications. Beginning in 1813, Dupin's call for teaching new geometry to researchers, students, and workers modeled later efforts to bring modern geometry into the textbook literature. His contributions more closely aligned with what would become differential geometry than projective geometry, but since this distinction did not yet exist, it would be artificial to remove Dupin from a study of modern geometry. Nevertheless, in the interest of space, I will leave aside a more technical discussion of his texts.[24]

In the introduction to his *Développements de Géométrie, avec des Applications à la stabilité des Vaisseaux, aux Déblais et Remblais, au Défilement, à l'Optique, etc.*, Dupin called for new concepts in elementary treatises. He intended his elemen-

---

[22] Vient ensuite l'application de l'algèbre à la géométrie; cette branche, due entièrement aux modernes, et dont la découverte leur a bientôt donné une immense supériorité sur les anciens, devait nécessairement changer de forme à mesure qu'elle s'étendoit et se perfectionnoit.

[23] As the history of complex numbers in the nineteenth century indicates, imaginary numbers held an ambiguous status within mathematics, and geometry in particular, through the 1820s (Flament 1997; Schubring 2005).

[24] For additional historical analyses of Dupin's contributions, see Christen-Lécuyer and Vatin (2009), and Bradley (2012).

tary treatise to serve as a sequel to the descriptive and analytic geometry introduced by Monge, most famously in *Géométrie descriptive* and *Application de l'Analyse à la Géométrie à l'usage de l'École Impériale Polytechnique* (Monge 1798, 1807, 1795). To accomplish this, *Développements de Géométrie* appeared in two parts, "Théorie" and "Applications" published, respectively, in 1813 and 1822 (Dupin 1813, 1822). He had studied descriptive geometry at the École Polytechnique with Monge, to whom he dedicated his text, and by the time the first part appeared, Dupin was already an acclaimed engineer and mathematician. Dupin described his work as written for "les élèves de l'École Polytechnique, ou des corps du Génie" (Dupin 1813, p. viii). Yet, while he declared his work a textbook, at the same time he promised to introduce new research.

> The progress of science is not truly fruitful, except when it also leads to the progress of elementary Treatises; it is through these writings that new concepts, reserved first for a small number of superior minds, finally becomes general knowledge, and extends its benefits into all parts that wait only for an intelligent application. (ibid, p. vii)[25]

In particular, Dupin promised to include results derived between 1805 and 1807, some of which had been previously published in the *Correspondance sur l'École Polytechnique*. Dupin further signaled his awareness of recent developments in geometry by summarizing the contributions contemporary geometers, and in particular former polytechniciens. Most of all, Dupin credited Monge and Carnot, who in turn provided a positive review of the book. Their recommendation, written along with Poisson on behalf of the *Académie des sciences*, was printed as a further introduction.

Dupin distinguished this book from his earlier articles, in that the treatment here would be simpler. The reviewers echoed this sentiment, acknowledging that Dupin contributed to both research and public works and pointed to "remarkably simple" new discoveries.

> The research that we are going to present proves that in the midst of the work with which he has been charged, M. Dupin has not lost sight of the objects of his first studies. It makes us wish that an engineer who reunites such extensive knowledge in geometry and analysis, would soon publish the work in which he proposes to apply them to questions of practice and public utility. (Dupin 1813, p. xx)[26]

The reviewers saw this enterprise as reflecting the founding goals of the École Polytechnique. Indeed, the entire "Théorie" text reflects the balance between writing for beginning students and experienced researchers. On the one hand, Dupin

---

[25] Les progrès de la science ne sont vraiment fructueux, que quand ils amènent aussi le progrès des Traités élémentaires; c'est par ces écrits que les conceptions nouvelles, réservées d'abord au petit nombre des esprits supérieurs, deviennent enfin des connaissances générales, et ramifient leurs bienfaits dans toutes les parties qui n'attendent qu'une application intelligente.

[26] Les recherches que nous venons d'exposer prouvent qu'au milieu des travaux dont il a été chargé, M. Dupin n'a pas perdu de vue les objets de ses premières études. Elles font desirer qu'un Ingénieur qui réunit des connaissances si étendues en géométrie et en analyse, publie bientôt l'ouvrage dans lequel il se propose de les appliquer à des questions de pratique et d'utilité publique.

occasionally apologized for providing too many details in a very elementary treatment.

> Perhaps, despite this, people well-versed in considerations of Geometry, will find still that I entered into too many details; but if these developments make that which seems too elementary easier, they will certainly not be superfluous for all the readers. (ibid, p. 25)[27]

On the other hand, Dupin at times chose his methods in order to maintain the practicality desired by engineers.

> If we only wrote for Geometers, we would have freed this latter part from all infinitesimal considerations; but in following the beautiful methods of the author of Fonctions Analytiques, it would have been less easy; and that ease is above all what we would like to be able to make possible, in order to generalize the study of theories truly useful to Engineers. (ibid, p. 68)[28]

These sentiments suggest a growing separation between professions in France, despite the goals of the École Polytechnique and Dupin's own contributions to both engineering and geometry. This distance seemed even more apparent by 1822 when his *Applications de Géométrie et de Méchanique, à la marine, aux ponts et chaussées, etc., pour faire suite aux Développements de Géométrie* appeared. Despite the many concrete applications within *Développements de géométrie* Dupin explained in his introduction to *Applications* that the first text had presented "abstract truths" that were "without practical utility" (Dupin 1822, p. xx). This sequel, which presumably could be read independently of the prefatory theory, would not be subject to "the same judgment." Though Dupin's endeavor to write at once for researchers and students was not emulated, he was joined in his commitment to introducing new geometry at the elementary level.

Garnier

Jean Guillaume Garnier (1766–1840) published the first edition of *Elémens de géométrie analytique* in 1808 as a "Traité que j'offre aux élèves" (Garnier 1808, p. iv). Garnier identified himself on the title page as an "Ancien Professeur à l'École Polytechnique, et Instituteur, à Paris"—indeed, he had been an assistant to Lagrange's courses between 1798 and 1802. By 1808, he was a teacher of transcendental mathematics at a lycée in Rouen. Moreover, as noted on the back cover, Garnier had published other textbooks on arithmetic, algebra, elementary geometry, statics, and differential and integral calculus. In the introduction to his first edition,

[27] Peut-être, malgré cela, les personnes très-versées dans les considérations de la Géométrie, trouveront-elles encore que je suis entré dans trop de détails; mais si ces développements rendent plus facile ce qui leur semblera trop élémentaire, ils ne seront certainement pas superflus pour tous les lecteurs.

[28] Si nous n'écrivions que pour des Géomètres, nous aurions pu dégager cette dernière partie de toute considération infinitésimale; mais en le faisant d'après les belles méthodes de l'auteur des Fonctions Analy-tiques, nous aurions été moins faciles; et c'est surtout ce que nous voudrions pouvoir être le plus possible, afin de généraliser l'étude des théories vraiment utiles à des Ingénieurs.

he credited the work of many other textbook authors associated with his former, prestigious, institution including Lacroix, Prony, Biot, Lefrançois, Boucharlat, Dinet, Puissant, Monge, Hachette, and Poisson.

In 1813, Garnier published a second edition, under a slightly different title, *Géométrie analytique ou application de l'algèbre à la géométrie*. He explained the need for this new edition by harshly criticizing his first edition.

> The first Edition of this Work lacks method, and consequently that which forms the principal merit of an elementary book: it desired several formulas which, without being exclusively preferable to others, advantageously replace them in the solution of a great number of questions; several solutions are not complete or thorough enough, others are diffcult; the problems of space are mixed with problems of two dimensions; finally the notation is often defective. (Garnier 1813, p. v)[29]

Garnier described this new treatise as "plus méthodique, plus soigné et plus complet" and credited particularly "les précieux matériaux" from Gergonne's Annales as well as the geometry research of L'Huilier and Puissant. As will be seen in the following section, Garnier included objects from modern geometry among this "precious material." Garnier was certainly familiar with Gergonne's *Annales* as he had submitted a brief article to the journal, which was published in 1813. Though most of Garnier's writings around 1813 were for textbooks, he would later contribute many brief articles to his own journal, *Correspondance mathématique et physique* (1825—1839), in almost every domain of pure and applied mathematics.

Biot

Garnier's inclusion of modern mathematics in 1813 demonstrates an exceptionally early adoption of certain recently published research. While Jean Baptiste Biot (1774–1862) thanked Garnier in the preface of his 1813 *Essai de Géométrie Analytique, appliquée aux courbes et aux surfaces du second ordre* (Biot 1813), not until the sixth edition, ten years later, did he also begin to include some of this same new content. The first edition of Biot's textbook appeared in 1802, written for prospective École Polytechnique students.

> This work is principally destined for the young people who are studying to enter the École Polytechnique. It results from lessons that I gave at the École Centrale de l'Oise. (Biot 1802, p. i)[30]

Part of Biot's qualifications included his own experience as a student at the École Polytechnique, where he enrolled in 1794. By 1803, he was a member of the *Institut*

[29] La première Édition de cet Ouvrage manque de méthode, et conséquemment de ce qui fait le principal mérite d'un livre élémentaire: elle laisse à désirer plusieurs formules qui, sans être exclusivement préférables à d'autres, les remplacent avantageusement dans la résolution d'un grand nombre de questions; quelques solutions ne sont pas complètes ou assez approfondies, d'autres sont pénibles; les problèmes de l'espace sont mêlés avec les problèmes à deux dimensions; enfin la notation est souvent défectueuse.

[30] Cet ouvrage est principalement destiné aux jeunes gens qui étudient pour entrer à l'École Polytechnique. Il est résultat des leçons que j'ai données à l'École Centrale de l'Oise.

de France, a professor of mathematics and physics at the *Collége du France*, and a professor of Astronomy at the *Faculté des Sciences* de Paris. The textbook appears to have been popular, as the next four editions quickly followed over the next ten years without many changes from the original volume. In the preface to his 1823 edition Biot apologized for his long hiatus, explaining understandably that he was prevented by other "occupations plus obligées, ou plus attrayantes" (Biot 1823, p. vii).[31] Even more than Garnier, Biot is connected to Parisian mathematics. Nevertheless, like Garnier, he credited the *Annales*, published in Nîmes and not formally connected to Parisian mathematical activity, for the new geometry he included in this edition.[32]

> I also believed I must no longer pass over in silence the properties of poles and polar lines first considered by Monge, and to which authors of the *Annales de Mathématiques* have given such elegant analytic developments. (Biot 1823, p. vii)[33]

Biot also cited Lagrange, Lacroix, and his brother-in-law Brisson for other modifications to his treatment of curves in this volume. In these numerous citations, Biot established a broad base of support for his new contents.

Vincent

The market for preparing future École Polytechnique students also included teaching elementary geometry. Alexandre Vincent (1797–1868) had been a student at the *École Normale* between 1816 and 1820, and first wrote an elementary geometry textbook dedicated to his students at the Collége royal de Reims in 1826. As he noted in the subtitle to the first edition, the *Cours de Géométrie Élémentaire* was "à l'usage des élèves qui se destinent à l'école Polytechnique ou aux écoles militaires." Vincent highlighted the pedagogical improvements to his approach, including distinct placement of practice problems and the statement of propositions without reference to lettered figures. He also announced additional material for strong students:

> For the rest, the things which are not indispensable are printed in small type, one could leave them aside, or reserve them as exercises for the strongest students. (Vincent 1826, p. iv)[34]

[31] Biot utilizes the exact same preface for his subsequent 1826 edition, which is identical to the fifth edition except for minor typographical corrections.

[32] In Otero (1997), Mario Otero statistically analyzes the distribution of content in Gergonne's *Annales* and finds that geometry was overrepresented as compared to other contemporary research publications.

[33] J'ai cru aussi devoir ne plus passer sous silence les propriétés des pôles et des lignes polaires considérées d'abord par Monge, et auxquelles les auteurs des Annales de Mathématiques ont donné des développemens analytiques si élégans.

[34] Au reste, les choses qui ne sont pas indispensables étant imprimées en petit caractère, on pourra les laisser de côté, ou les réserver comme exercices pour les élèves les plus forts.

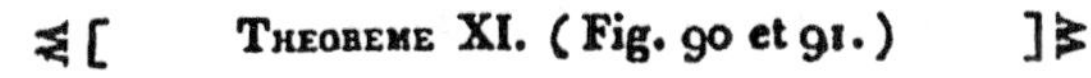

Fig. 4.1 Vincent's notation in Vincent (1832)

The use of small font enabled those who wanted to focus on only the entrance exam material to skip these sections. With respect to content, Vincent credited Lacroix, Francoeur, Legendre, Dupin, Develey, and Gergonne "dont j'ai plus d'une fois consulté les intéressantes annales" (ibid, p. v).

Vincent published a second edition in 1832 based on feedback from other teachers, a review by Augustin Cournot in Lycée, and "un rapport très étendu adressé par M. Ampère au Conseil royal de l'instruction publique" (Vincent 1832, p. v). Many of these changes were organizational, such as better in-text references to corresponding problems. Vincent eliminated the use of small font, due to complaints about legibility. However, rather than deleting the challenging content, he added more, highlighting the elementary principal properties of transversals, radical axes, and poles and polars. In this edition, Vincent denoted the extracurricular status of this material with using the symbol of a left and right facing sideways M (Fig. 4.1).

> In addition, I have noted these theories, like several others, as well as a great number of propositions which one does not require students to prepare for exams, by an ostensible sign that advertises to the reader in a hurry to arrive at the goal, and lacking necessary time or volition to explore in detail the numerous avenues of the science of extension, that he can pass over without being subsequently required to retrace his steps. (ibid, p. xi)[35]

Since geometry exams did not contain new content, any modern geometry could only be included as supplementary in this genre of textbook. As in 1826, Vincent acknowledged a large number of his contemporaries, here also adding Bergery, Terquem, and finally "des Annales de Mathématiques, dont le savant rédacteur a eu l'obligeance de m'adresser en outre diverses spécialement appropriées à mon ouvrage" (ibid, p. xv). In fact, Vincent wrote three articles for Gergonne's *Annales* between 1825 and 1826 though none of these were on subjects of elementary geometry.

Didiez

While Garnier, Biot, and Vincent composed geometry textbooks connected to Paris, and more particularly the École Polytechnique, outside this mathematical center authors also took opportunities to write textbooks with modern content.

Even so, N. J. Didiez is the only one of the authors in this corpus apparently without ties to the École Polytechnique. Little is known today of Didiez beyond his

[35] Au surplus, j'ai noté ces théories, comme plusieurs autres, ainsi qu'un grand nombre de propositions que l'on n'exige point des élèves qui se présentent aux examens, d'une signe ostensible qui avertira le lecteur pressé d'arriver au but, et manquant du temps ou de la volonté nécessaire pour explorer en détail les avenues nombreuses de la science de l'étendue, qu'il peut passer outre sans se trouver exposé par la suite à revenir sur ses pas.

published books and their reviews. He published the first part of his *Cours Complet du Géométrie* on planar elementary geometry in 1828, advertising this text as the first in a four-part series that would progress through three-dimensional elementary geometry, planar analytic geometry, and finally three-dimensional analytic geometry (Didiez 1828). These books represented an ongoing private mathematics course that Didiez had been teaching for the past eight years. The course and associated texts are described in an *Annales* review of Didiez's volume on arithmetic, published in 1825.

> M. Didiez has been giving public mathematics courses in Paris for several years. Preferring to surrender to his own ideas than to subject himself to follow those of another, but wanting to avoid the loss of time which the dictation of lesson entails, he proposes to publish a simple summary of his lessons; and it is the summary of those of arithmetic that he presents today. (Gergonne 1826a)[36]

Following these geometry texts, Didiez promised a subsequent series on applications to the "arts d'imitation et de construction" (Didiez 1828, p. i). However, if any of the other anticipated volumes ever appeared, there are no longer any publicly available extant copies. Though the circumstances of publication might indicate a less established author, Didiez's geometry textbook was published by Bachelier with drawings engraved by Adam—both well-respected individuals in the textbook medium.

Didiez dedicated his book to Dupin in a very elaborate full-page spread that listed the latter's many accomplishments: Dupin's membership at the Institut de France, various public honors, and position at the Conservatoire des Arts et Métiers. By 1828, following an initiative by Dupin in Paris, after-hours courses for workers were widely available in many French metropolitan areas. Likewise, Didiez may have decided to offer his own courses in the evening in order to attract a wide range of students and employed persons.

Bergery

Claude Lucien Bergery (1787–1863) more directly emulated Dupin by spearheading the public education efforts in Metz, which resulted in his books *Cours de sciences industrielles. Géométrie appliquée à l'industrie* (Bergery 1825), 1826).[37]

Bergery categorized his subject as "géométrie pratique." Practical geometry, as defined by François Joseph Servois in *Solutions peu connues de différens problèmes de géométrie-pratiques* concerned the study of executing "diverse geometric operations on the terrain" (Servois 1803, p. 1). Bergery devoted his introduction to addressing "les ouvriers et artistes" from Metz and explained that he had wanted to

---

[36] M. Didiez fait à Paris, depuis plusieurs années des cours publics de mathématiques. Aimant mieux s'abandonner à ses propres idées que de s'astreindre à suivre celles d'autrui, mais voulant éviter la perte de temps qu'entraîne la dictée des cahiers, il se propose de publier un simple résumé de ses leçons; et c'est le résumé de celles d'arithmétique qu'il présente aujourd'hui.

[37] See Vatin (2007) for a scientific biography of Bergery.

teach such a course since 1821 but only with the movement toward public education initiated to Dupin had such intentions been realized.[38] Bergery elaborated the practical potential of an education in geometry.

> The one Geometry has three distinct branches: the geometry of the straight line and circle, whose use is daily; that of curves, which explains many wonders; descriptive geometry, which one can call the language of constructions, and which applies to architecture properly speaking, to stone cutting, to carpentry, to painting, to sculpture and to a great number of other arts. (Bergery 1825, p. xix)[39]

The first volume was intended for students with only a basic knowledge of arithmetic. Bergery followed with a "second part" on the geometry of curves applied to industry, published in 1826 (Bergery 1826).[40]

Bergery enrolled at the École Polytechnique in 1806 and had taught geometry and engineering at the École royale de l'artillerie in Metz since 1817. There he worked with Poncelet, whom he mentioned as another instructor in the first edition of 1825. By Bergery's 1828 second edition, Poncelet appears as a primary influence (Bergery 1828a). Bergery framed this new edition as providing the necessary prerequisites for a young geometer to study "sans peine, les Propriétés projectives des figures dans le bel ouvrage de M. Poncelet, et de s'élever à des connaissances qui, jusqu'à présent, ont été rangées dans la Géométrie transcendante" (Bergery 1828a, p. vii).

Nevertheless, Bergery departed from his colleague on certain issues of simplicity, generality, and vocabulary, which will be explored in the following section. Bergery promised the most clear, methodical, and complete volume of practical geometry. Achieving this required a balance between accessible and comprehensive scope. Bergery refrained from too much technical language:

> I have abstained from several scientific expressions which, in the end, teach nothing, and each time that I have been obliged to employ them, I have taken care to explain them by equivalent expressions taken from common language. (ibid, p. xx)[41]

---

[38] Dupin's efforts toward public education within the context of engineering are discussed in Grattan-Guinness (1984), particularly Sect. 8.

[39] La seule Géométrie a trois branches distinctes: la géométrie de la ligne droite et du cercle, dont l'usage est journalier; celles des courbes, qui explique tant de merveilles; la géométrie descriptive, qu'on peut appeler la langue des constructions, et qui s'applique à l'architecture proprement dite, à la coupe des pierres, à la charpenterie, à la peinture, à la sculpture et à un grand nombre d'autres arts.

[40] This brief second volume, *Cours de Sciences Industrielles. Seconde Partie. Géométrie des courbes appliquée à l'industrie* covers the properties and construction of conic sections, lemniscates, spirals, cycloids, and a wide variety of other curves and analogous surfaces. Bergery directed the reader interested in "demonstrations de ceux de principes que nous avons seulement énoncés" to the *Annales*, Poncelet's *Traité* Brianchon's *Mémoire sur les lignes du second ordre*, among other contemporary texts. However, none of the modern geometry contained in these suggested readings is in this volume except in the form of succinctly stated results, where any modern techniques were obscured.

[41] Je me suis abstenu de plusieurs expressions scientifiques qui, dans le fond, n'apprennent rien, et chaque fois que j'ai été obligé d'en employer, j'ai en soin de les expliquer par des équivalens pris dans le langage vulgaire.

By contrast, Bergery argued that it was necessary to introduce the recently discovered objects and expressions from contemporary geometry research into elementary geometry books.

> For some time I have regretted not finding in elementary books any notion of Transversals which make the practice of geometry so simple, Poles and Polars, conjugate Points, radical Axes, similitude Centers, Centers of gravity, and traces, rather frequently used, many of which result from recently discovered principles. Why, in effect, not try to place these new riches from science at the door of practitioners who can use them daily? (ibid, pp. vi—vii)[42]

Thus, this second edition marked a radical departure from his earlier texts in terms of new content and contemporary references. While Bergery wrote for a local audience, his text achieved wide distribution throughout France as well as at bookstores as far as Liége and London. A third edition, which retained the modern geometry from the second, appeared in 1835.

Terquem

Each of the aforementioned authors demonstrated considerably more reliance on contemporary research articles—particularly those in the *Annales*—than their fellow textbook writers. Nevertheless, Olry Terquem (1782—1862) eclipsed them all in his efforts to connect elementary geometry to modern geometry. Terquem published his *Manuel de géométrie* in 1829 for "l'usage des personnes privées des secours d'un maître" (Terquem 1829, p. i). The book opened with a two-column page of authors cited alphabetically from Anonyme to Vincent. These names ranged in time and fame from Archimedes to Durrande (a young geometer, who had published several articles in elementary geometry in Gergonne's *Annales* before his death in his early 20s). Gergonne is the most widely represented, with four listed citations.

Terquem was a teacher of mathematics and the librarian at the Dépôt Central de l'Artillerie Paris, but he is perhaps better known for co-founding the *Nouvelles Annales de Mathématiques* with Gerono in 1842. Like his manuals from a decade before, in this journal Terquem would strive to engage young geometers in new research and by many accounts succeeded. As observed by Chasles in an obituary from 1863,

> These *Nouvelles Annales*, in the modest format of 1 in octavo and a moderate price, were destined especially for teachers and numerous candidates to the Écoles of Government: Écoles Normale, Polytechnique, Militaire, de Marine, etc. M. Terquem, in exciting young geometers about research on posed questions, and welcoming their attempts, in making

[42] Depuis quelque temps on regrettait de ne trouver dans les livres élémentaires aucune notion sur les Transversales qui rendent si simple la pratique de la Géométrie, sur les Pôles et les Polaires, sur les Points conjugués, sur les Axes radicaux, sur les Centres de similitude, sur les Centres de gravité, et sur des tracés, d'un usage assez fréquent, dont plusieurs résultent de principes récemment découverts. Pourquoi, en effet, ne pay essayer de mettre ces nouvelles richesses de la science à la portée des praticiens qui peuvent s'en servir tous les jours?

> them aware of new facts of science, either by this publications or by his individual communications, rendered a great service to mathematical studies. (Chasles 1863, p. 245)[43]

Terquem's *Manuel de géométrie* aimed for a similar audience, but in the format of a compact textbook. In this book, Terquem criticized the standard curriculum in which students progressed slowly from elementary geometry, to planar and spherical trigonometry, to conic sections, to second-degree surfaces, and finally to projective procedures and descriptive geometry. Many students dropped out along the way, even though for physical science and industrial arts "les propriétés des sections coniques, les moyens graphiques sont au moins aussi importans à connaître que la mesure des distances, des aires, des volumes, but ordinaire de la géométrie élémentaire" (Terquem 1829, p. iv).

To correct this omission, Terquem proposed a one-year geometry course that would blend elementary, analytic, and descriptive geometry into a single subject accessible to any student with a previous course in algebra.[44] Along with condensing several years of geometry, Terquem intended to blend the writings of ancient texts with contemporary geometers:

> We have applied ourselves to editing this Manual following the ideas just given, remaining in the limits prescribed to this nature of work; we have given all that is essential in the ancient treatises and in the writings of contemporary geometers. (ibid, p. vi)[45]

This was not an idle promise. Of all the texts in our corpus, Terquem's is the most closely correlated with the methods and directions in recent research publications. Rather than a result of centralized administration, individual innovation drove the use of modern geometry in these seven titles. This personal initiative marked other forms of nineteenth-century education. In *Espaces de l'enseignement scientifique et technique*, historians d'Enfert and Virginie Fonteneau describe the potential for the individual in the evolution of teaching.

> One such approach leads equally to consider the relations between the individuals and the institutions within which they evolve, between individual actions and collective enterprises. The questions then concern constraints of the environment where these individuals exert their action as well as the margins of movement or the possible options available to them. For a number of actors evoked in this work, the realization of their projects or those, which

[43] Ces Nouvelles Annales, dans le modeste format de 1 in octavo et d'un prix modéré, étaient destinés surtout aux professeurs et aux nombreux candidats aux Écoles du Gouvernement: Écoles Normale, Poly-technique, Militaire, de Marine, etc. M. Terquem, en excitant les jeunes géomètres à des recherches sur des questions proposées, en accueillant leurs essais, en les tenant au courant des faits nouveaux de la science, tant par cette publication que par ses communications individuelles, rendait un grand service aux études mathématiques.

[44] Of historiographical interest, Terquem noted that he would not be straying too far from the "method of the ancients" since Euclid was essentially using algebra "sans signes, mais en phrases" in "five of his fifteen books" (iv). Terquem's reference to fifteen books of Euclid indicates that he was working from a different manuscript tradition than his contemporaries. For instance, François Peyrard divided Euclid's *Elements* into twelve books in his French translation (Peyrard 1804).

[45] On s'est appliqué à rédiger ce Manuel d'après les idées qu'on vient d'émettre, se tenant dans les limites prescrites à cette nature d'ouvrages; on a donné tout ce qu'il y a d'essentiel dans les anciens traités et dans les écrits des géomètres contemporains.

> they had been assigned is not exempt from personal interests in terms of career, status and social recognition. (D'Enfert and Fonteneau 2011, p. 11)[46]

For Garnier, Biot, and Vincent, the content of their textbooks was prescribed by the course of study at the École Polytechnique, and the inclusion of modern geometry required circumventing these prescriptions. In this aspect, elementary and analytic geometry appear equally conservative. Bergery and Terquem were less institutionally bound, but nevertheless utilized their introductions to justify the inclusion of newer concepts as providing practical shortcuts for students. Whether in introductions or citations, each of these authors indicated their knowledge of contemporary geometry research, often through specific articles published in the *Annales*. However, the majority viewpoint as expressed by great names like Lacroix and Legendre indicates that knowing about modern geometry was necessary, but not sufficient, for including modern geometry in a textbook. Authors also had to believe that there were educational advantages to these innovations.

The following section will examine what objects from modern geometry were perceived as worth importing and the contexts in which they were employed. These decisions reveal how geometers attempted to resolve tensions between theory and application and to strike a delicate balance between stating general principles and practicing specific constructions.

5 New Objects

New research in pure geometry coincided with new, specialized vocabulary. This trend is especially apparent in the *Annales des mathématiques pures et appliquées*, in which the terms pole and polar were first introduced and radical and ideal objects quickly proliferated (Servois 1810). Within these articles, the adoption of these terms signaled an awareness of contemporary results as well as a willingness to employ new results in further research. However, the use of an author's vocabulary did not necessarily coincide with support of his underlying method. Similarly in books, the new vocabulary of modern geometry could be adapted to more conservative contexts. Citations suggest that textbook authors were also aware of diverse contemporary approaches, and deliberately chose definitions that could benefit a pedagogical setting.

In particular, this section will focus on the concepts of pole and polar and then radical and similitude through the introduction of these objects between authors and

[46] Une telle approche conduit également à considérer les relations entre les individus et les institutions au sein desquelles ils évoluent, entre les actions individuelles et les entreprises collectives. Les interrogations portent alors sur les contraintes du milieu où ces individus exercent leur action ainsi que sur les marges de manoeuvre ou les possibilités de choix dont ils disposent. Pour nombre d'acteurs évoqués dans cet ouvrage, la réalisation de leurs projets ou de ceux qui leur ont été assignés n'est d'ailleurs pas exempte d'intérêts personnels en terme de carrière, de statut et de reconnaissance sociale.

editions. Finally, the treatment of imaginary points in the work of some authors reflected a willingness to extend teaching to the forefront of research.

Together these authors demonstrate common strategies for incorporating new material into textbooks. The following cases will demonstrate that geometric objects entered the textbook literature when they could be adapted to multiple contexts and represented through simple constructive language and figures.

5.1 Poles and Polars

The pole of a line was first defined by Servois in his 1810 solution to a posed problem published in the first volume of the *Annales*. His definition comprises the opening paragraph to his article:

> A line and a second degree curve being given, I call pole of the line, a point in the plane of this line and the curve around which turn all the chords of contact points of pairs of tangents to the curve from different points on the line: (Servois 1810, p. 338)[47]

The fact that such a point uniquely existed was often attributed to Monge, who had not assigned any special name to this property. In 1810, Servois simply stated and then applied the definition to finding a triangle circumscribing a given curve. In the third volume of the *Annales*, Gergonne introduced the corresponding polar of a point and proved the existence of the pole and polar in terms of coordinate equations (Gergonne 1813). Following Poncelet's publications on polar reciprocity in 1822 and Gergonne's use of duality in 1826, pole and polar were often associated with dual relationships between definitions and theorems (Poncelet 1822; Gergonne 1826b). Textbook authors' introductions to pole and polar can be classified under three strategies, which can be summarized as copy and paste, constructive innovations, and new applications and properties.

Copy and Paste

In an 1812 article on finding the distance between the centers of circles inscribed and circumscribed to a given triangle, Garnier promised a new edition of his *Application de l'algèbre à la géométrie*, which appeared the following year (Garnier 1812, p. 347). Both the article and his new textbook demonstrated a confident literacy in recent research published in the *Annales*, and, indeed, the addition of poles and polars in Garnier's second edition marked the influence of the new journal.

Garnier included poles and polars in a section on problems concerning tangents of second-degree curves. He began with a secant and a second-degree curve of the

---

[47]Une droite et une ligne du second ordre étant assignées, j'appelle pôle de la droite, le point du plan de cette droite et de la courbe autour duquel tournent toutes les cordes des points de contact des paires de tangentes à la courbe issues des différens points de la droite:

form, $ay^2 + cx^2dy + ex = 0$. Each secant passed through the curve in two points and tangent lines drawn from these two points would intersect in a point on the plane. Garnier then showed that when the secant turned around a given point G with coordinates $(g; h)$, then the locus of tangent intersection points was given by the equation

$$(2cg + e)x + (2ah + d)y + eg + dh = 0$$

a straight line. He added that "inversely" if the tangent intersection points lay on a straight line, then the corresponding secants all passed through the same point.

Following this theorem, Garnier provided a definition.

> Because of the relation, which exists between the point G and the line which is the locus of vertices of circumscribed angles, this point has been called the pole of this line, and one calls the line the polar of the point G. (Garnier 1813, p. 165)[48]

Though Garnier later used the result relating secants and tangents, this definition was the only mention of pole and polar in his text. In fact, Garnier's exposition was nearly identical to that of Gergonne from the *Annales* in 1813 down to the use of coefficients and the concluding paragraph. For comparison,

> Because of the relation which exists between the point (P) and the line (Q), this point has been called the Pole of this line; and one can, inversely, call the line (Q) the Polar of the point (P). (Gergonne 1813, p. 297)[49]

Undoubtedly, Garnier took his descriptions of poles and polars from this article, or perhaps a previous unpublished version, though Gergonne was not directly cited. While this might seem like plagiarism today, the complete appropriation of Gergonne's proof aligns with the acknowledged lack of originality and infrequent citations in textbook writing of the early nineteenth century. More surprisingly, Garnier here exhibited a remarkably quick publication process with the ability to integrate the previous year's newest results and vocabulary. The novelty of pole and polar at this time may also explain why they did not appear elsewhere in Garnier's text, despite their obvious abbreviating power.

Constructive Innovations

Garnier offered little innovation, except possibly introducing pole and polar to a new audience. By contrast, ten years later Biot used his analytic geometry textbook to provide a more constructive, graphical treatment of poles and polars. First, Biot defined pole and polar with respect to a circle centered at a point C as shown in his figure 43 (see Fig. 4.2). He proposed that the definition could be extended by analogy to all second-order curves.

---

[48] A cause de la relation qui existe entre le point G et la droite qui est le lieu des sommets des angles circonscrits, ce point a été appelé le pôle de cette droite, et on peut appeler la droite la polaire du point G.

[49] A cause de la relation qui existe entre le point (P) et la droite (Q), ce point a été appelé le Pôle de cette droite; et on peut, à l'inverse, appeler la droite (Q) la Polaire du point (P).

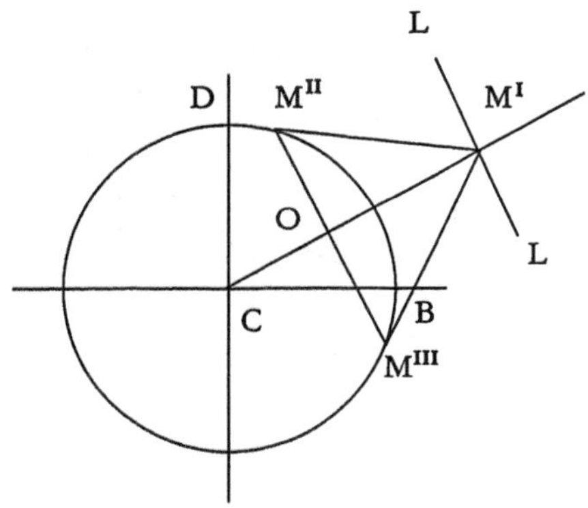

Fig. 4.2 Figure 43 in (Biot 1823)

> As analogous properties are found in all second order curves, one employs abbreviated denominations to express them. The point where the chords meet is called the pole of the line LL, from where the tangents are drawn, and reciprocally; this line is called the polar line of the point O. (Biot 1823, pp. 197–198)[50]

This quote might seem to suggest that Biot would simply rely on his circle construction and analogy, but as the text progressed he systematically provided constructions for the pole and polar of an ellipse (figure 58) and a parabola (figure 76), and only applied analogy to the case of the hyperbola (see Fig. 4.3).

> In continuing to follow, in calculations, the analogy between two curves, we will arrive at a similar construction, to determine the line which contains the vertices of the pairs of tangents, when one knows the intersection of the chords, and reciprocally. The similitude is so perfect that there is no need to explain here the application of this method, and it will suffice, to realize it, to cast ones' eyes on fig. 91. (ibid, p. 307)[51]

Earlier articles in pure geometry included constructions of poles and polars for the circle, but Biot's specific constructions for the ellipse and parabola appear unique. His constructions emphasized a different property of poles and polars than that stated in the definition as can be seen in the case of finding the polar of a point O with respect to a parabola.

Beginning with a parabola and a coplanar point O, draw a line OM parallel to the axis of the parabola and meeting the curve at the point M. Through the point M draw a tangent TMT′ to the curve. Then draw the chord M″M‴ through the point O and parallel to TMT′, where M‴ is the point where the chord meets the parabola. From

[50] Comme des propriétés analogues se retrouvent dans toutes les lignes du second ordre, on a employé des dénominations abrégées pour les exprimer. Le point où l cordes concourent, s'appelle le pôle de la droite LL, d'où les tangentes sont menées, et réciproquement; cette droite se nomme la ligne polaire du point O.

[51] En continuant de suivre, dans les calculs, l'analogie des deux courbes, on arrivera à une construction pareille, pour déterminer la droite qui contient les sommets des couples de tangentes, quand on connaîtra le point de concours des cordes, et réciproquement. La similitude est si parfaite, qu'il n'est pas besoin d'expliquer ici l'application de cette méthode, et qu'il suffira, pour s'en rendre compte, de jeter les yeux sur la fig. 91.

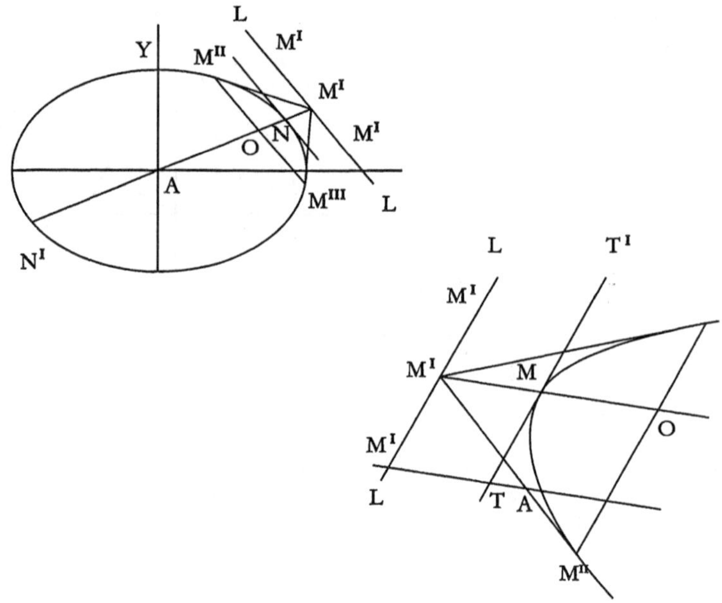

Fig. 4.3 Figures 58 and 76 in (Biot 1823)

the point M″ draw a new tangent that will meet the line OM in the point M′. Then the polar is the line through M′ and parallel to TT′, which is drawn as LM′L.

This construction is fairly simple to execute as it only involved drawing two tangents. Since Biot was writing a textbook in analytic geometry, his commitment to demonstrating particular cases is striking. Indeed, he proved each construction analytically using the specific equation of the given second-order curve rather than a general second-order curve. Biot also made notice of exceptional cases, such as if the point O lays on the axis of the parabola.

New Applications and Properties

Without the use of coordinate equations, Bergery and Vincent limited their exposition to the case of the circle. Nevertheless, these authors reveal the wide-ranging potential of objects from modern geometry to play various roles within the elementary geometry context.[52]

---

[52]Within his chapter on "Des systèmes qu'on peut former sur un plan avec trois lignes droites, ou circulaires" Didiez showed that:

130. When two lines are tangent to the same circumference, if one imagines that the intersection point of these tangents moves along a straight line drawn arbitrarily through this point, the tangent lines and points will change position; this will be the same for the chord of contact, but in

Bergery used the vocabulary of pole and polar, stating a defining property. He explained that the name pole "signified pivot of rotation" and then provided the construction of a polar using two chords and their associated tangents, which he had just shown how to construct in an earlier section (Bergery 1828a). To find the pole, one could either follow a similar construction of tangents or use the fact that the pole of a line lies on the perpendicular drawn from the center of the circle to the given line.

Bergery followed his constructions with practical applications. First, he described how to use poles and polars in a pivoting physical model that could produce a circular movement from a rectilinear movement without the use of gears (ibid., pp. 137–138). Bergery's applications show the potential benefit of poles and polars beyond theoretical geometry. Further, his construction demonstrated that these new objects were no more difficult to find than tangent lines to circles.

An alternative construction can be found in the second edition of Vincent's textbook (Vincent 1832). Vincent introduced pole and polar in an optional section on the propreties of transversal lines. For a given line OA and coplanar point P not on the line, one could construct lines through the point that meet the given line at points A; B; C; D; .... Then from the point O, any transversal to these new lines would meet them at points, respectively, denoted a; b; c; d; .... Each of the pairs of diagonals Ab and aB, Bc and bC, Cd and cD, ... will meet at a point, p; q; r; ... and the geometric locus of these points was a line through the point O. This line Op was the polar of the point P with respect to the angle AOa and reciprocally, P was the pole of the line Op.

Though two intersecting lines form a degenerate case of a conic section, Vincent's construction of pole and polar without an obvious curve was unusual. He then considered a circle centered at O with radius OA. If on the line OA, one took two points P; Q on the same side of the circle's center such that the product of their distances to the circle's center equaled the distance OA^2, then these points would be conjugate to each other with respect to the given circle.

Finally, if one drew a perpendicular through P or Q to the line PQ, this perpendicular would be the polar of the other conjugate point. That is, perpendicular PM was the polar of Q and perpendicular QN is the polar of P with respect to the circle

---

all the positions that the latter can take, it will not stop passing through the same point situated on a line drawn from the center of the circumference, perpendicular to the direction according to which one moves the intersection points of the tangents. (Didiez (1828), 124)

130. Lorsque deux droites sont tangentes à une même circonférence, si l'on conçoit que le point de concours de ces tangentes se meuve le long d'une droite menée comme on voudra par ce point, les tangentes et les points de contact changeront de position; il en sera de même de la corde de contact, mais dans toutes les positions que cette dernière pourra prendre, elle ne cessera de passer par un même point situé sur une droite menée du centre de la circonférence, perpendiculairement à la direction suivant laquelle on fait mouvoir le point de concours des tangentes.

Didiez's proof relied on the fact that the radius of the circle would be the mean proportional between the distance from the center to the chord of contact and the distance from the center to the tangents' intersection point. Though Didiez seemed comfortable with new vocabulary, he did not use the terms pole and polar here and this result concluded the section on "The system formed by two straight lines and a circle" without further applications.

Fig. 4.4 Illustration of Vincent's pole and polar construction

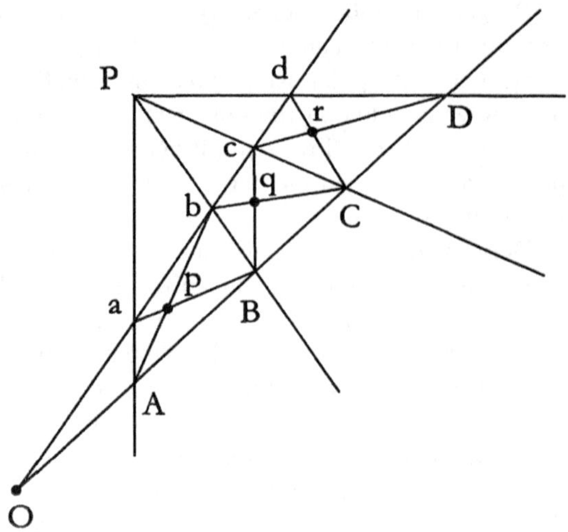

OA. In parentheses, Vincent noted that the names were based on the fact that these points and lines have analogous properties with respect to the circle as the above defined poles and polars have with respect to the given angle (see Fig. 4.4).

Having shown the relationship between transversal lines, conjugate points, and poles and polars, Vincent proceeded to prove several theorems about properties of poles and polars. Many of these theorems appeared in reciprocal pairs, such as the pole of all lines through a given point is on the polar of this point and reciprocally the polar of all points on a given line passes through the pole of this line. These proofs mainly utilized the fixed product definition of conjugate points discussed above. The theorems culminated in a pair of theorems (now known as Pascal's and Brianchon's theorems)—that for all hexagons inscribed to a circle, the points of intersection of opposite sides taken two by two are collinear, and reciprocally for circumscribed hexagons. Vincent followed these by several corollaries on quadrilaterals and triangles. He concluded with a parenthetical reference.

> See the *Annales de Mathématiques* in various places, and notably volume XIV, page 39 and following.—See also the *Correspondance sur l'École Polytechnique*. (Vincent 1832, p. 216)[53]

Indeed, Vincent's theorems summarized many recent results from these two publications including articles by Brianchon, Gergonne, and Poncelet.

As writers of analytic geometry textbooks, both Garnier and Biot would have expected their readers to be familiar with elementary geometry from a previous course, while Didiez, Bergery, and Vincent only anticipated their students knew elementary arithmetic. In his geometry textbook, Terquem attempted to create a volume that introduced beginning students to elementary and analytic geometry at

[53] Voyez les *Annales de Mathématiques* en divers endroits, et notamment Tome XIV, page 39 et suiv.– Voyez aussi la *Correspondance sur l'École Polytechnique*.

the same time. This meant a lengthier text with a combination of purely constructive and coordinate equation definitions and examples. Terquem's definition of poles and polars is computationally similar to Bergery's, emphasizing the turning property of the chords. Terquem then proved the validity of this property by computing with the harmonic ratios created by the intercepted segments.

Terquem stated that the intersection of two given polar lines is the pole of the line containing their poles with respect to the same circle. He claimed this property was "easy to prove" and proceeded by showing how it could be applied in problem solving. For instance, he employed poles and polars in showing how to find a third point such that its distance to two given points is in a given ratio (Terquem 1829, p. 147).

Though Terquem initially defined poles and polars with respect to circles, he later concluded that "the polar properties of circles [...] belong to second degree curves," which he argued by considering the poles and polars as "the angular projections of analogous lines and points situated in a circle." This projective relationship furnished "an easy means to draw a tangent to a second degree curve through a point not on the curve" (ibid, p. 350). Terquem explained the consequences of this relationship in particular cases, such as when the pole of a diameter is at infinity and that the directrix of a hyperbola is the polar of its closest focus.

Finally, in a note to a discussion of the volume of ellipsoids and elliptic paraboloids, Terquem generalized polar reciprocity with the use of coordinate equations.

> If from a point A lying on a curve of degree p, one draws tangents to a curve of order m, the tangent points are situated on a line of order m–1 (13.); each position of the point A responds to another curve of tangent points; all are tangent to a curve of degree
> $(m + p - 2)^2$. In the particular case where $p = 1$, this latter curve reduces to $(m-1)^2$ points through the tangent curves constantly pass. This proposition with its reciprocal contains the general theory of polar curves. (ibid, p. 444)[54]

Terquem located this result at the very end of his text, which served to show his familiarity with the general theory of polar lines without alienating his intended audience of beginners. He was exceptionally generous with citations to contemporaries in the majority of his book, but did not provide any references for poles and polars. This may also reflect his knowledge and involvement with research mathematics, where the concepts were simply part of the standard lexicon by the end of the 1820s.

The use of poles and polars in textbooks reveal a range of methods for incorporating objects from modern geometry. Garnier represents one end of the spectrum, excerpting the treatment of poles and polars directly from Gergonne's article without any significant modification, commentary, or applications. In contrast, both Biot and Bergery provided more practically oriented texts by showing the visual and

[54] Si d'un point A situé sur une ligne du degré p, on mène des tangentes à une ligne de l'ordre m, les points de contact sont situés sur une ligne de l'ordre m–1 (13.); chaque position du point A répond à une autre courbe de points de contact; toutes sont tangentes à une courbe du degré $(m + p - 2)^2$. Dans le cas particulier où $p = 1$, cette dernière ligne se réduit à $(m - 1)^2$ points par lesquels passent constamment les courbes de contact. Cette proposition avec sa réciproque renferme la théorie générale des lignes polaires.

concrete properties. Biot presented precise constructions for almost all cases of conics, rather than simply giving a generic or circle-based construction. Bergery followed his definition of poles and polars with many examples of potential practical applications, oriented toward physical tools of measurement and design. Finally, Vincent and Terquem summarized recently proved theorems and solved problems that exhibited the significant role played by pole and polar in the past two decades. Since Vincent professed to be writing this section for advanced students, he could also introduce them to the latest research and even suggest ways in which they might contribute.

Though Garnier showed that textbooks could simply copy and paste from research articles, the latter presentations suggest ways in which textbook literature instead provided more nuanced understanding of poles and polars, in drawing connections to practical considerations and in synthesizing accumulated results to show the current state of knowledge.

5.2 *Centers and Axes*

Like poles and polars, the concepts of similitude and radical defined relationships between points and lines with respect to other coplanar figures.

Similitude Without Radicals

In Friedelmeyer's history of transformations in the nineteenth century, he explains how similitude was generalized from polygons to general curves at the end of the eighteenth century "soit par une traduction analytique, soit par une mise en relation d'éléments homologues" (Friedelmeyer 2016, p. 22). Euler introduced the "similitude center" in *De centro similitudinis*, but most early nineteenth century French authors attributed similitude centers and axes to Monge (pp. 24–27). For instance, Monge provided a brief account of similitude in an article published in the *Correspondance sur l'École Polytechnique* in 1814 (Monge 1814). In this two-page article, he showed how to calculate the coordinates of a similitude center for any two second-degree curves of the form

$$Ax^2 + By^2 + Cxy + 2Dx + 3Ey - 1 = 0$$

He concluded by referencing the *Traité des surfaces du second degré* by Monge and Hachette. In this text, the authors proved that when second-degree surfaces are cut by parallel planes, any two sections are similar and similarly placed curves, and so can be considered as parallel sections of a conic surface (Hachette and Monge 1813). While similarity is an important subject of this book, the expression similitude does not appear. The lack of systematic vocabulary makes the concept of similitude somewhat difficult to trace.

For instance, in his chapter on "Systems that one can form on a plane with three straight or circular lines," Didiez provided an extensive discussion of similarity, similitude, and homology for triangles. Following French convention, he attributed the similitude center of two similar and similarly placed triangles to Monge.

> This intersection point of the three lines drawn through the homologous vertices of two similar and similarly placed triangles has been named by MONGE, the similitude center of two triangles. It is the direct similitude center or the inverse similitude center, according to whether the two triangles are directly or inversely similar. (Didiez 1828, p. 88)[55]

However, he surprisingly referenced a specific volume and page number in the *Annales* for his definition of the similitude axis (p. 89). Though Didiez did not mention the author, the article is Gergonne's 1827 interpretation of Steiner's 1826 article on circle tangency first published in Crelle's *Journal für die reine und angewandte mathematik* (Steiner and Gergonne 1827). While this article contains much new vocabulary including radical axes, Gergonne made no claim to originality in the use of similitude axes. Further, Didiez restricted this initial definition to triangles, and Gergonne, in the cited text, defined similitude centers and axes for general polygons and circles. The citation thus appears merely as a jumping off point for Didiez, who modified the scope and order to suit the prominent role of triangles in his text.

Only in subsequent chapters did Didiez define similitude centers of axes for arcs of circles (Didiez 1828, p. 132) and then similar polygons (p. 179), each with references back to his initial triangle definition.[56]

In his study of circles, Didiez explained that the point of contact between two tangent circles would "evidently" be a direct or inverse similitude center depending on the kind of tangency. He then used this property to solve the problem of describing a circle passing through a given point A and tangent to two given circles on a plane. This question has four solutions, which utilized properties of similitude centers, as can be seen in the case where the tangent circles are exterior.

> Suppose the question is solved, and AED is a circumference passing through the point A and exteriorly tangent at D and E to the given circumferences. The tangent points D and E will be the inverse similitude centers of the circumferences to which they belong (no. 142). (Didiez 1828, p. 204)[57]

---

[55] Ce point de concours des trois droites menées par les sommets homologues, de deux triangles semblables et semblablement situés, a été nommé par MONGE, le centre de similitude des deux triangles. Il est centre de similitude directe ou centre de similitude inverse, suivant que les deux triangles sont directement ou inversement semblables.

[56] A nearly identical definition of similitude can be found in the second edition of Étienne Bobillier's Cours de Géométrie from 1834. Based on the table of contents from the 1832 edition (only available at Archives Départamentales de la Marne, and thus not included in our corpus), little changed between the first and second edition. In both of these editions, Bobillier's text is very brief (less than 100 pages in the first edition) and similitude is the only concept from modern geometry adopted in these first two editions. For more on Bobillier and this text, see dos Santos (2015).

[57] Supposons la question résolue, et soit AED une circonférence passant par le point A et touchant extérieurement en D et E les circonférences données. Les points de contact D et E seront les centres de similitude inverse des circonférences auxquelles ils appartiennent (no. 142).

In turn, these solutions form the basis of how Didiez solved the Apollonius problem, one of the most famous geometry problems of nineteenth-century geometry. Didiez had introduced similitude with triangles, but he found the largest scope of application in considering similitude between circles.

Uniting Similitude with Radicals

The history of radical axes and centers is less ambiguous. Louis Gaultier first defined radical axes and centers for two and three given circles in the *Journal de l'École Polytechnique* in 1813 (Gaultier 1813). Like similitude, these radical objects proliferated through geometry research articles by the 1820s (for instance, Steiner and Gergonne 1827; Plücker 1826; Bobillier 1827). Although these centers and axes were introduced roughly contemporaneously with poles and polars and appeared in the same journals, they were even less frequently used in textbooks—neither appears in the books of Didiez, Garnier, or Biot. The earliest instance that I found of these objects is in textbooks from the mid-1820s.

Vincent included similitude centers for polygons in the optional content of both his 1826 and 1832 editions. His definition was essentially the same as the one given by Didiez (except using "internal" for "inverse" and "external" for "direct"), which he extended to tetrahedra and then general polyhedra in a later chapter. In the second edition, Vincent significantly expanded and updated this material.

First, Vincent incorporated recent publications. In his discussion of polygons, he mentioned the "série de propositions sur les figures semblables, nouvellement démontrées par M. Chasles" though without an exact citation (Vincent 1832, p. 179). Further, he extended the concept of similitude from his first edition. By considering circles as regular polygons, Vincent determined there would be two similitude centers for any pair of circles. This definition could be applied to all possible cases of circle position (internal, external, tangent, concentric) as well as the degenerate cases where one of the given circles was a straight line or a point. In a later section, Vincent showed that the three centers of similitude of three similar and parallel polygons would lie on a straight line, the similitude axis. He then considered the case of three circles, which would have three internal and three external similitude centers, which determine four similitude axes. Here, too, Vincent demonstrated his knowledge of contemporary articles:

> These axes are the only common homologous lines that the three polygons or three circles can create. (See the *Annales de Mathématiques* of M. Gergonne, Volume XIII, page 197.) (ibid, 206)[58]

This is an article on tangent circles written anonymously as a letter to the editor of the journal, but subsequently attributed to Gergonne (1823).

[58] Ces axes sont les seules droites homologues communes que puisent avoir les trois polygones ou les trois cercles. (Voy. les *Annales de Mathématiques* de M. Gergonne, Tome XIII, page 197.)

Vincent adopted the concept of radical in his second edition, defining a radical axis as the locus of points from which one can draw tangent lines of equal length to two given circles. In examining particular cases, Vincent concluded that two concentric circles would not have a radical axis since there are no points from which tangents of equal length can be drawn. He defined the radical center of three given circles as the point of intersection of their three radical axes.

The use of sideways M symbols to set off each of these results gives the impression of disjoint results. Thus when Vincent followed this definition with a theorem on similitude centers, he appeared to be changing topics. He proved that when one drew two secants through the similitude center of two given circles, O; O′, the eight resulting points of intersection could be taken four by four to define four new circumferences. Vincent called these four new circles the reciprocal circles to O, O′ relative to their similitude center. With this new concept, he then returned to radical axes in a corollary, revealing that each similitude center of two circles was the radical center of all their reciprocal circles relative to this similitude center. Thus, the new objects emerged as interdependent, consequently strengthening the relative importance of similitude in this second edition.

All of the problems are located at the end of Vincent's text, where he applied these new objects from modern geometry in several constructions, such as finding a circle tangent to three given circles. For this problem, Vincent provided three solutions. The first did not invoke any modern geometry. The second employed both similitude centers and radical axes, drawing on their common properties through reciprocal circles. Vincent explained that this second solution was superior to the first.

> Apart from the exceptional case that we have just signaled, the second construction has the great advantage of being applicable, when it is conveniently modified, to problems that were solved following number 276 inclusive. (Vincent 1832, p. 327)[59]

These earlier problems that Vincent referenced were versions of finding a circle subject to three conditions including passing through a given point or being tangent to a given line. Finally, Vincent attributed his third solution, which employed similitude axes and poles, to Gergonne in volume XVII of the *Annales*. This was the same article cited earlier by Didiez, Gergonne's interpretation of Steiner (Steiner and Gergonne 1827). For more on this problem, Vincent recommended the recent text of Bergery.

> We also encourage students to consult the Géométrie of M. Bergery. They will find there the discussion of different cases that can lead to the second construction, which, moreover, is due to M. Poncelet. (Vincent 1832, p. 328)[60]

---

[59] En mettant à part le cas d'exception que nous venons de signaler, la deuxième construction a le grand avantage de pouvoir s'appliquer, lorsqu'elle est convenablement modifiée, aux problèmes qui ont été résolus depuis le numéro 276 inclusivement.

[60] Nous engageons aussi les élèves à consulter la Géométrie de M. Bergery. Ils y trouveront la discussion des divers cas que peut présenter la deuxième construction que l'on doit d'ailleurs à M. Poncelet.

This citation suggests that textbook innovation could be contagious, or at least that these authors were aware of the novelties in each other's work. Rather than starting with triangles, Bergery first defined similitude for circles. He introduced similitude in a section on drawing secants, using proportions between segments to show that "the common secants to two circles, determined by parallel radii, have two points of intersection A; A' which are always conjugated to each other, in such a manner that the ratio of the two parts formed by each of these points, on the line of the centers, is equal to the ratio of the radii" (Bergery 1828a, p. 129). After explaining the designations "direct" and "inverse," Bergery promised that "we will see the basis of these denominations, when we will study similar polygons," which formed the topic of a later section (p. 130).

In an investigation of intersecting circles, Bergery applied similitude to the study of radical circles, axes, and centers. For a given circle and a coplanar point A, if one drew a secant intersecting the circle at D and C and passing through its center, then the circle centered at A with radius equal to the mean proportional between AC and AD would be radical to the first (p. 150). Bergery explicitly limited his study to radical circles where the point A lay outside of the given circle. Then the locus of centers of all circles radical to two given circles would be their radical axis, and analogously three given circles would define the radical center.

Returning to similitude, Bergery found that one could always "describe a circle that cuts two others A; B into four points C; D; E; F where they are met by two secants through one of their centers of similitude, as long as the four points are not on parallel radii" (p. 151). This new circle, Bergery named a reciprocal circle on account of the relationship between a similitude center I and the four points, namely ID: IC:: IF: IE. Combining all of these new terms, he concluded this section in showing that "one or the other of the similitude centers of two circles A; B, is the radical center of all reciprocal circles relative to this similitude center" (p. 152). This result, using slightly different vocabulary, had only just been published in the research articles of Steiner and Plücker (Steiner 1826, Plücker 1827). Bergery did not reference these geometers, but later credited Poncelet's solution to finding a circle tangent in the same way to three given circles. The "elegant construction due to M. Poncelet" utilized similitude centers and axes, radical centers and axes, and poles (Bergery 1828a, p. 162).

Having demonstrated how to construct tangent circles, Bergery emphasized their practical applications.

> Drawing tangent circles is frequently used in the construction of machines; gears that fit with other gears, or pinions, or lantern gears rest on these drawings. [...] Tangent circles also form the curves that workers call ovals when they are completed or closed, and anses de panier when they are only halves. (ibid., p. 174)[61]

[61] Le tracé des cercles qui se touchent, est d'un usage fréquent dans la construction des machines: c'est sur ce tracé que repose celui des roues dentées qui engrènent soit avec d'autres roues dentées, soit avec des pignons, soit avec des lanternes. [...] Ce sont aussi des cercles tangens les uns aux autres qui forment ces courbes que les ouvriers nomment ovales quand elles sont complètes ou fermées, et anses de panier lorsqu'une moitié manque.

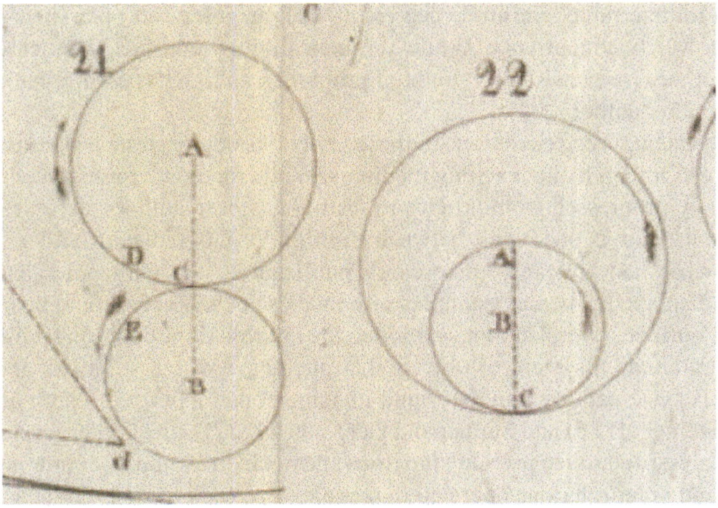

Fig. 4.5 Figures 21 and 22 in (Bergery 1828a), planche VI

Bergery illustrated these applications in more detail with several helpful figures. In figures 21 and 22 (Fig. 4.5), he used arrows to portray how the rotation of externally tangent circles will result in opposite motion, while internal tangency will create rotations in the same direction. In figure 24, Bergery drew *anses de panier* using the arcs of tangent circles. Bergery developed numerous methods for constructing several different kinds and sizes of arcs and ovals using tangent circles. These figures could achieve both aesthetic appeal and architectural utility. Bergery was thus able to connect some of the most recent research in planar geometry to the anticipated practices of his working students.

Like Didiez, Vincent, and Bergery, Terquem also attended to new findings from contemporary sources. However, he created some of his own new vocabulary (which did not catch on) to replace the term radical axis with dishomologous lines in order to emphasize its homologous relationship to the two given circles (Terquem 1829).

Terquem first used similitude in the context of similar polygons, defining the similitude center as the intersection of lines that pass through the homologous vertices between two similar polygons (p. 168). After showing the use of similitude in constructing similar polygons, he extended the concept to similar curves since "the polygons ABCDE, abcde are similar, and have the same similitude center as the similar curves in which they are inscribed" (p. 173). Analogously, he defined similitude centers in his study of similar polyhedra (p. 271).

Terquem applied these centers and axes toward finding similar and tangent curves. For instance, he demonstrated that the diagonals of a hexagon circumscribed to a conic section intersected in the same point, a property he attributed to Brianchon, and, as noted above, Vincent also included (p. 227). Terquem limited his proof to the case of the circle, which he attributed to another *Annales* writer, M. Durrande. He constructed a series of circles centered around the vertices of the circumscribed hexagon to determine similitude centers as well as poles and polars that could be applied to verify concurrence. While Terquem slightly modified Gaultier's vocabulary, he otherwise consistently credited properties and proofs as originating in the works of other authors.

The frequent, precise citations in these textbooks demonstrate a significant overlap between research and teaching mediums. While the flow of referenced information usually progressed from articles to textbooks, these authors also extended the applications and connections between similitude and radical centers and axes beyond what had appeared in research publications. As with poles and polars, Bergery displayed concrete examples of how these new objects could be applied to industry. Further, several authors exploited the interrelations between similitude and radicals to create a system of objects that interacted within solutions and proofs. While this systematicity may be found in some contemporary research articles — such as Steiner (1826) and Steiner and Gergonne (1827)—the longer format of textbooks enabled authors to present numerous and detailed examples of how similitude, radical, and polar relations operated in tandem.

The adaptability and abbreviating power of these objects to different formats may explain why they were attractive to a diversity of textbook writers. For instance, as similitude could be introduced with respect to polygons or circles, these concepts could be adjusted to suit the order of a textbook that introduced circles early in a Euclidean style or defined them later as the limit of regular polygons. Since many of these textbooks were not first editions, this meant that authors did not have to dramatically reformat their texts in order to incorporate aspects of modern geometry.

The selection, organization, and presentation of these objects from modern geometry exemplify a "process of elementarization" described by Schubring as "the transposition of knowledge into teachable knowledge and a related method" (Schubring 1987, p. 47). Knowing the elements of a subject could serve to develop deeper and more advanced research. However, not all of modern geometry was easily "elementarized" in this time period.

5.3 Imaginaries

Research publications in modern pure geometry—-particularly those of Poncelet, Brianchon, and Chasles—-did not shy from interpreting the imaginary points and lines that emerged from the application of algebra to coordinate equations (Poncelet 1822; Brianchon 1817; Chasles 1828).[62] Yet in the first third of the nineteenth century even textbooks on analytic geometry only gingerly treated imaginary objects.

[62] The question of representing imaginary points in geometry was a different one than that solved by the complex plane and also debated in Gergonne's *Annales* in this same time period.

Briefly overviewing the presence and absence of imaginaries in the above texts underlines a deep ambivalence in large part driven by concerns for constructive practicality.

A footnote by Dupin on the language of geometry serves to situate the inherent limitations of imaginary numbers in geometry textbooks.

> Often, in transcendental geometry, when one considers extension in all degrees of generality, one must speak at times of points, curves, surfaces, volumes. In order to avoid this long enumeration, we thought it necessary to designate all the magnitudes by the general expression graphic magnitudes, that is to say, capable of being drawn. (Dupin 1813, p. 15)[63]

The understanding that geometric quantities must be graphic and therefore drawable created a firm distinction between the real and the imaginary. In early nineteenth-century geometry, imaginary points, lines, surfaces, etc. could not be figured. Further, there were no attempts within these textbooks to assign an ontological status to imaginary objects. Nevertheless, the expression "imaginary" could play one of several roles within textbooks.

Imaginary values entered geometry textbooks through square roots. Once introduced, they could be discarded or studied. To take a common example, in using coordinate equations to represent second-degree curves, the geometers Dupin, Garnier, Didiez, and Terquem derived one real and one imaginary diameter or axis of a hyperbola. Unlike the real diameter, the imaginary diameter did not intersect the hyperbola. Garnier showed that if the half-diameter of a hyperbola, A', was real then the conjugate half-diameter B' would be of the form $B'\sqrt{-1}$. While the $\sqrt{-1}$ indicated that the points of intersection would be imaginary, Garnier employed the real B′ coefficient to show that the difference of squares of conjugate demi-diameters was equal to the difference of squares of the demi-axes (Garnier 1813, p. 148). Thus, the quantitative value of the imaginary diameter continued to display information about the curve. Including imaginary conjugate diameters also reinforced a general treatment of conic sections without exceptional cases.

The use of real coefficients allowed imaginary diameters to serve a function in better understanding properties of conic sections. Imaginary diameters thus functioned as an extension of imaginary points of intersection, when two curves shared an imaginary point and no real points, geometers concluded that the curves did not intersect. Yet calculations that resulted in imaginary values were also read as indicating a lack of existence (Garnier 1813, p. 75) or an impossible situation (Terquem 1829, p. 439). To give a sense of the language, consider how Biot introduced the square roots of negative numbers as impossible roots.

> Finally, if B extends past A, the circle described by the point C as center, with A as radius, will never cut the indefinite line AB. The points X; X', thus cannot be found in this circum-

[63] Souvent, dans la géométrie transcendante, où l'on considère l'étendue dans tous les degrés de généralité, on doit parler à la fois de points, de lignes, de surfaces, de volumes. C'est pour éviter cette longue énumération, que nous avons cru devoir désigner toutes ces grandeurs par l'expression générale de grandeurs graphiques, c'est-à-dire, susceptibles d'être figurées.

> stance, and so the solution of the proposed question will be impossible. This is also what the equation between the numerical values shows; because, if b extends past a, the radical part $\sqrt{a^2 - b^2}$, which is common to the two roots, becomes imaginary, and consequently, the two roots are impossible. (Biot 1823, p. 19)[64]

This is also illustrated in Garnier's "Example II" where he showed that the parabola $x^2 + yx = 0$ was imaginary when $y > 1/4$. This part of the parabola was invisible.

Though less common, imaginary solutions also emerged in elementary geometry. For instance, in finding a circle of given radius that is tangent to a given line and passes through a given point, Vincent noted the possible cases: "There will be two solutions which can reduce to one only or become imaginary" (Vincent 1832, p. 312).[65] Vincent did not explicitly define what an imaginary solution was, and only used the term in the context of finding points—in this case the center of the desired circle. He more generally described geometric problems that do not lead to a real constructive solution as "impossible" such as in one of the cases of constructing a circle tangent to a given circle and a given line and passing through a given point: "Finally, the problem is impossible when the point is interior and the line is exterior to the given circle" (ibid, 321).[66] By contrast, Vincent also explained for exactly which configurations a construction would be possible, thus implying other cases were not.

To varying degrees the expressions: does not intersect, no longer exists, and becomes impossible, served to convey the non-constructive status of imaginary values. Given the practical constructive aims of these textbooks, it is not surprising that Poncelet's ideal chords comprised of imaginary points were not adapted for beginning students in the same way as radical and similitude axes. Even Poncelet's admitted admirers, like Bergery and Terquem, or those who used ideals in their own research articles, like Bobillier, continued to dismiss imaginary solutions (Bobillier 1828).

6 Reception

Though these texts cracked the traditional textbook mold, the majority was not received as particularly groundbreaking. Some acknowledged reception could be found in reviews published in the *Bulletin*. The review of Vincent comments on his decision to place problems at the end of the text, rather than interspersed with the

[64] Enfin, si *B* surpassait *A*, le cercle décrit du point *C* comme centre, avec *A* pour rayon, ne couperait pas du tout la droite indéfinie *AB*. Les points *X*; *X'*, ne pourraient donc pas s'obtenir dans cette circonstance, et ainsi la solution de la question proposée serait impossible. C'est aussi ce que l'équation entre les valeurs numériques montre; car, si b surpasse a, la partie radicale $\sqrt{a^2 - b^2}$, qui est commune aux deux racines, devient imaginaire, et conséquemment, les deux racines sont impossibles.

[65] Il y aura deux solutions qui pourront se réduire à une seule ou devenir imaginaires.

[66] Enfin, le problème est impossible quand le point est. intérieur et le droite extérieure au cercle donné.

definitions and theorems (Anonymous 1827b, p. 82). Didiez was praised for his care and zeal and recommended to "teachers and students" (Anonymous 1828, p. 321). Though both reviewers summarized the contents of the respective textbooks, the authors' uses of modern geometry went unmentioned. Terquem's reviewer (signed D.—which from the list of contributors suggests either Duhamel or Dupin) noted the author's ambitious plan to combine elementary geometry, rectilinear and spherical trigonometry, conic sections, second-degree surfaces, and descriptive geometry into "un petit volume" (D. 1829, p. 1). However, the review focused primarily on Terquem's citations to contemporary geometers, even offering a correction to attribute a proof to Lacroix rather than Querret. Certainly, attention to geometers' new theorems and proofs is one of the unusual features of Terquem's textbook, but it did not reveal much of the book's contents. Similarly, in reviewing the first edition of Vincent's *Cours de géométrie élémentaire* for *Le Lycée, Journal de l'instruction rédigé par une société de professeurs, d'anciens élèves de l'Ecole normale*, Cournot mentioned that the author "distinguishes, by a very small typeface, less essential theories" without elaborating what these theories were (Cournot 2010, p. 551).

By contrast, Dupin's textbooks appeared to be quite influential through the 1820s, particularly for his unified approach to theory and practice. In an article based wholly on Dupin's *Developpements*, entitled "Démonstration des principaux Théorèmes de M. Dupin sur la courbure des surfaces," Gergonne proposed to introduce Dupin's work to a wider audience:

> We initially dreamed to give a simple analysis of the work of M. Dupin; but, this task having already been accomplished by several journals, we thought to do something more convenient and more useful simultaneously, in presenting here the principal points of the doctrine of the author rather briefly in order to enable its introduction in elementary treatises, where its importance must henceforth be found. (Gergonne 1814, p. 368)[67]

Thus, Dupin's success in obtaining a research article for his textbook may be attributed to his multiplatform publication approach, which combined articles, presentations to the Institut des sciences, and reviews in diverse publications (a strategy Poncelet also utilized to gain readers for his *Traité*).

Dupin's commitment to education also inspired the subsequent pedagogical literature of both Didiez and Bergery.[68] Indeed, in a self-review and defense written for the *Société des lettres, sciences, arts et agriculture*, Bergery framed his textbook as following Dupin's commitment to bring new results to elementary geometry and a more inclusive audience.

> M. Ch. DUPIN, in creating the Courses of industrial science, has opened the methodical and logical path; I dared to enlarge it and push it further; but authors who have preceded and followed us, remain in the narrow paths of the workshop routine; believing that workers are

[67] Nous avions d'abord songé à donner une simple analise de l'ouvrage de M. Dupin; mais, cette tâche ayant déjà été remplie par plusieurs journaux, nous avons pensé faire une chose plus convenable et plus utile à la fois, en présentant ici les principaux points de la doctrine de l'auteur dans un cadre assez resserré pour qu'il soit permis de l'introduire dans les traités élémentaires, où son importance doit désormais lui faire trouver place.

[68] Though neither geometer included the indicatrix or conjugate tangents in their own textbooks.

> ignorant of the simplest facts, they are dedicated to describe them more or less well, without the least attempt to explain them. March thus and you will arrive, they have said; as for why, you do not need it; we know it for you, that is enough. (Bergery 1828b, p. 20)[69]

Bergery claimed to include the new theory behind these practices, but faced criticism in a review by fellow textbook author Francoeur that appeared in the *Revue Encyclopédique*. Though overall impressed with Bergery's text, Francoeur doubted that the readers would benefit from Bergery's more advanced treatment:

> But we do not see that this geometry is more appropriate for teaching this class of men than that for all types of students; and, except for the choice of examples, which are in effect appropriate for industry, the work could also be well placed in the hands of all genres of readers. It appears to me that a geometry for artisans must be a simple collection of propositions, clarified by easy demonstrations, when that is possible, and by numerous applications to the arts. [...] We do not know how it is more useful to the student to teach him the succession of truths which compose an elementary treatise than to clearly conceive the details, and how these details themselves, when they are too numerous, are detrimental to the general instruction, that we want to give. (Francoeur 1828b, pp. 753–754)[70]

In his defense, Bergery countered that the theory enabled viewing the connection between various results and actually served to attract the interest of workers toward science:

These are not the geometric laws of nature that we would like to see disappear; these eternal applications which reveal a supreme intelligence, greatly excite the interest of workers and are very well suited to create love of science, so their suppression is not an evil. (ibid., p. 18).[71]

Bergery described his elementary treatise as the most extended and fruitful yet written. Though Francoeur had not singled out poles, polars, radicals, or similitude in his review, these aspects of modern geometry might similarly be criticized as outside the domain of useful pedagogical instruction. Francoeur's own textbook,

---

[69] M. Ch. DUPIN, en créant les Cours de sciences industrielles, a ouvert la voie méthodique et logique; j'ai osé l'élargir et la pousser plus avant; mais des auteurs qui nous ont précédés ou suivis, se sont plus à rester dans les étroits sentiers de la routine des ateliers; croyant les faits les plus simples ignorés des ouvriers, ils se sont attachés à les décrire plus ou moins bien, sans chercher le moins de monde à les expliquer. Marchez ainsi et vous arriverez, ont-ils dit; quant au pourquoi, vous n'en avez pas besoin; nous le savons pour vous, cela suffit.

[70] Mais on ne voit pas que cette géométrie soit plus propre à l'enseignement de cette classe d'hommes qu'à celui de toute espèce d'étudians; et, sauf le choix des exemples, qui sont en effet appropriés à l'industrie, l'ouvrage pourrait tout aussi bien être mis entre les mains de tous les genres de lecteurs. Il me paraît qu'une géométrie pour les artisans devrait être un simple recueil de propositions, éclairées par des démonstrations faciles, lorsque cela se peut, et par de nombreuses applications aux arts. [...] On ne sait pas assez combien il est plus utile à l'étudiant de lui faire saisir l'enchaînement des vérités qui composent un traité élémentaire, que d'en concevoir nettement les détails, et combien ces détails eux-mêmes, lorsqu'ils sont trop multipliés, nuisent à l'instruction générale, qu'on veut donner.

[71] Ce ne sont pas non plus les lois géométriques de la nature qu'on voudrait voir disparaître; ces applications éternelles qui révèlent une suprême intelligence, excitent trop l'intérêt des ouvriers et sont bien trop propres à faire aimer la science, pour que leur suppression ne soit pas un mal.

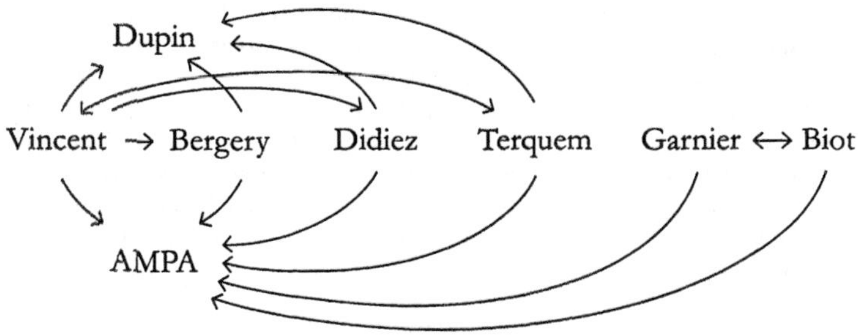

Fig. 4.6 Diagram of citations between textbooks

Cours complet de mathématiques pures, did not include any of these new objects (Francoeur 1828a).

Nevertheless, textbook authors could and did lead others to include modern geometry. As noted above, many of the first textbooks to promote modern geometry were second or subsequent editions, often revised following suggestions from colleagues and students. There was a substantial amount of citation among the authors in this study illustrated in the diagram in Fig. 4.6.

While Gergonne's *Annales* appears as a clear source of modern geometry, it is otherwise difficult to ascertain who borrowed from whom. Yet, looking slightly beyond the first third of the nineteenth century, there is some evidence of a ripple effect.

In the first two editions of his *Cours de Géométrie*, Étienne Bobillier (1798–1840) included the similitude center and axes with respect to similar polygons and polyhedra (Bobillier 1832, 1834). Though Bobillier was a frequent contributor to the *Annales* and utilized poles, polars, and ideal chords in this research, none of these objects were included in this course taught in Angers between 1831 and 1832 and published by the École Royale des arts et métiers of Châlons (Bobillier 1827, 1828).

However, the third edition of this text appeared in 1837 with an added dedication to A. Vincent (Bobillier 1837). Here, the content is much expanded and includes poles, polars, radicals, and the Apollonius problem. Though the treatment is not identical to Vincent's 1832 *Cours de géométrie élémentaire*, the overlap in scope is remarkable. It appears that Vincent's use of modern geometry inspired Bobillier in rewriting this edition.

But textbooks could also lose modern content. While Bobillier died in 1840, his text was subsequently adopted by the minister of agriculture and commerce for the Écoles nationales d'Arts et Métiers. Curiously, by the tenth edition (printed in 1850), most traces of the modern geometry, except for similitude, had disappeared and the text is much closer to the second than the third edition (Bobillier 1850). Likewise, in the third edition of Vincent's text, published in 1834, the author explained his decision to suppress the modern geometry that he had previously included in small font or offset by unique notation.

> The extension that I have given to several theories, having brought the volume of the preceding edition beyond the limits between that which one is accustomed to see included in

> elementary Geometry, I thought it necessary to suppress here the chapter on Transversals and Polars, the majority of the problems on Tangents, a chapter where I had very briefly shown the principles of the theory of Projections, and finally, a portion of the Numerical Problems, which are found multiplied beyond measure. One can consult, for the theory of Transversals, and that of Tangents, special works, notably those of Carnot, of MM. Brianchon, Poncelet, Gaultier de Tours, the *Annales de Mathématiques* of M. Gergonne, and finally, the Treatises of Geometry of MM. Bergery and Didiez, whose plan, less restricted than mine, admits developments which, for me, were nothing other than inconveniences. (Vincent 1834, p. v)[72]

The citations to Bergery and Didiez further emphasize the scarcity of textbooks containing modern geometry at this time. Vincent's reference to restraints implies that the text's role in preparing students for entrance exams may have curtailed the inclusion of new objects. The case of Bobillier's tenth edition suggests a similar institutional oversight and limitation.

7 Conclusion

In 1810, when Gergonne began publishing his *Annales*, he discussed the many functions and advantages of a journal devoted to mathematics.

> [...] a periodical that allows Geometers to establish a commerce among themselves or, to put it better, a kind of community of views and ideas; a periodical that spares them from vainly engaging in research already undertaken by others; a periodical which guarantees to each the priority of the new results that they come across; a periodical finally, which assures everyone's work publicity, not less honourable for them than useful to the progress of science. (Gergonne 1810, pp. i—ii)[73]

This public exchange of new ideas aimed toward scientific progress provides a contrast to the slow repetition characteristic of most geometry textbooks, even though pedagogical goals extended to journal publications as well. Gergonne introduced

---

[72] L'extension que j'avais donnée à plusieurs théories, ayant porté le volume de l'édition précédente au-delà des limites entre les quelles on est accoutumé à voir renfermer la Géométrie élémentaire, j'ai cru devoir supprimer dans celui-ci, le chapitre des Transversales et des Polaires, la plus grande partie des problèmes sur les Contacts, un chapitre où j'avais très brièvement exposé les principes de la théorie des Projections, et enfin, une portion des Problèmes Numériques, qui se trouvaient multipliés outre mesure. On pourra consulter, pour la théorie des Transversales et celle des Contacts, les ouvrages spéciaux, notamment ceux de Carnot, de MM. Brianchon, Poncelet, Gaultier de Tours, les *Annales de Mathématiques* de M. Gergonne, et enfin, les Traités de Géométrie de MM. Bergery et Didiez, dont le plan, moins restreint que le mien, admettait des développemens qui, pour moi, n'étaient pas sans inconvéniens.

[73] [...] un recueil qui permette aux Géomètres d'établir entre eux un commerce ou, pour mieux dire, une sorte de communauté de vues et d'idées; un recueil qui leur épargne les recherches dans lesquelles ils ne s'engagent que trop souvent en pure perte, faute de savoir que déjà elles ont été entreprises; un recueil qui garantisse à chacun la priorité des résultats nouveaux auxquels il parvient; un recueil enfin qui assure aux travaux de tous une publicité non moins honorable pour eux qu'utile au progrès de la science."

the *Annales* as above all consecrated to "recherches qui auront pour objet d'en perfectionner et d'en simplifier l'enseignement" (p. ii). However, while most articles in the *Annales* and similar journals could certainly be read by students or used by their instructors in creating teaching material, they were not explicitly presented by their authors as such. Instead, the research was framed as an end in itself, or to be used by other participants in the shaping "a community of views and ideas."

Without the community of readers and writers afforded by journals, books were self-sufficient by default. Book authors noted in their prefaces whether any arithmetic, algebra, or additional geometry might be required in advance, and if so, occasionally cited a few sources that might serve as preliminaries. Articles contained none of these explicit prerequisites, instead adopting intext references to cite particular concepts or results. Most books contained very few such references. Further, despite the prevalent redundancy, there is no evidence in the texts of opposition among textbook writers with respect to priority or potential plagiarism. Authors primarily restricted their particular criticisms to pedagogy and order of exposition.

Authors who participated in research and read articles seem to have been more likely to integrate new objects into their teachings, but due to institutional restrictions and limitations on student mathematical background, these instances remained rare and tentative. Textbook introductions demonstrated awareness of this novelty, by often highlighting material beyond the common curriculum. Even so, authors who included objects from modern geometry in some textbooks did not always continue to do so in later editions or other titles.

When authors adapted modern geometry to textbooks, they demonstrated careful consideration. First, the objects appeared as practical and constructive. Students received explicit instructions on finding and applying poles, polars, similitude, and radicals to solving problems from geometry, design, and engineering. The potential audience of mathematics textbooks mostly consisted of students who would not become mathematicians. Nevertheless, these students could find utility in certain concepts from recent research when presented in concrete terms.

Secondly, the length and summary nature of textbooks provided the possibility of bringing together systems of objects and displaying several methods for deriving results. While these objects and results were not original, textbook authors could curate an informative selection. Consequently, students could better observe the multiple potentials of these objects as tools through examples. No textbook claimed to be exhaustive in this respect, but even bringing together material from more than one source added value for the student interested in future research. The efficacy of these contributions is only speculative, however. It would be interesting, for example, to see whether Vincent's more advanced students actually pursued research in modern geometry at a greater rate than the average candidate for the École Polytechnique.

By the mid-twentieth century, many considered projective geometry as a teaching subject, with little research potential (Coolidge 1934, Bourbaki 1960). This article shows one aspect of this evolution—how textbooks began to represent modern geometry. This process began as research was still in flux, but notably poles, polars, similitude, and radicals featured in projective geometry textbooks a century later. Rather than signaling the ossification of modern geometry, these early manifestations for a student audience served to expand potential applications and contexts in an emergent discipline.

References

Anonymous. 1827a. 216. *Annales de mathématiques pures et appliquées*, par M. Gergonne; Tom. XVII, nos. 9 et 10, mars et avril 1827. *Bulletin des Sciences Mathématiques, Astronomiques, Physiques et Chimiques* 7: 273–279.

———. 1827b. Mathématiques élémentaires. *Cours de Géométrie élémentaire*; par M. Vincent. *Bulletin des Sciences Mathématiques, Astronomiques, Physiques et Chimiques* 7: 82–83.

———. 1828. Mathématiques élémentaires. *Cours de Géométrie*; par M. Didiez. *Bulletin des Sciences Mathématiques, Astronomiques, Physiques et Chimiques* 10: 322.

Barbin, Evelyne. 2015. Top-down: the role of the Classes Préparatoires aux Grandes Écoles in the French teaching of descriptive geometry. 1840–1910. In *"Dig where you stand" Proceedings of the third International Conference on the History of Mathematics Education*, ed. Johan Prytz, Kristin Bjarnadöttir, Fulvia Furinghetti, and Gert Schubring, 49–64. Uppsala: Department of Education Uppsala University.

Belhoste, Bruno. 1995. *Les Sciences dans l'Enseignement Secondaire Français: Textes Officiels*. Paris: Institut national de recherche pédagogique (INRP).

———. 1998. Pour une réévaluation du rôle de l'enseignement dans l'histoire des mathématiques. *Revue d'Histoire des Mathématiques* 4: 289–304.

Bergery, Claude-Lucien. 1825. *Géométrie Appliquée à l'Industrie à l'Usage des Artistes et des Ouvriers*. Metz: Lamort.

———. 1826. *Géométrie des Sciences Industrielles. Seconde Partie. Géométrie des Courbes Appliquée à l'Industrie*. Metz: Lamort.

———. 1828a. *Géométrie Appliquée à l'Industrie à l'Usage des Artistes et des Ouvriers*. 2nd ed. Metz: Mme. Vve. Thiel.

———. 1828b. Précis des travaux de la société par M. Bergery. *Société des Lettres, Sciences, Arts et Agriculture*: 17–20.

Bézout, Étienne, and Antoine-André-Louis Reynaud. 1812. *Cours de Mathématiques: À l'Usage de la Marine et de l'Artillerie*. Paris: Vve Courcier.

Bioesmat-Martagnon, Lise, éd. 2011. Éléments d'une Biographie de l'Espace Projectif. Nancy: Presses Universitaires de Nancy.

Biot, Jean-Baptiste. 1802. *Essai de Géométrie Analytique, Appliqué aux Courbes et Aux Surfaces du Second Ordre*. Paris: Crapelet.

———. 1810. *Essai de Géométrie Analytique, Appliqué aux Courbes et Aux Surfaces du Second Ordre*. 4th ed. Paris: J. Klostermann Fils.

———. 1813. *Essai de Géométrie Analytique, Appliqué aux Courbes et Aux Surfaces du Second Ordre*. 5th ed. Paris: J. Klostermann Fils.

———. 1823. *Essai de Géométrie Analytique, Appliqué Aux Courbes et Aux Surfaces du Second Ordre*. 6th ed. Paris: Bachelier.

Bobillier, Étienne. 1827. Géométrie de situation. Recherches sur les lois générales qui régissent les lignes et surfaces de tous les ordres. *Annales de Mathématiques Pures et Appliquées* 18: 157–166.

———. 1828. Géométrie de situation. Recherches sur les lois générales qui régissent les courbes algébriques. *Annales de Mathématiques Pures et Appliquées* 19: 106–114.

———. 1832. *Cours de Géométrie*. Châlons-sur-Marne: Cornet Paulus.
———. 1834. *Cours de Géométrie*. 2nd ed. Châlons-sur-Marne: Cornet Paulus.
———. 1837. *Cours de géométrie*. 3rd ed. Châlons-sur-Marne: Barbat.
———. 1850. *Cours de Géométrie*. 10th ed. Paris: Guiraudet et Jouaust.
Bossut, Charles. 1800. *Cours de Mathématiques*. 6th ed. Paris: Firmin Didot.
Bourbaki, Nicolas. 1960. *Éléments d'Histoire des Mathématiques*. Paris: Hermann.
Boyer, Carl B. 1956. *History of Analytic Geometry*. New York: Scripta Mathematica.
Bradley, Margaret. 2012. *Charles Dupin. (1784–1873) and His Influence on France: The Contributions of a Mathematician, Educator, Engineer, and Statesman*. Amherst: Cambria Press.
Brianchon, Charles Julien. 1817. *Mémoire sur les Lignes du Second Ordre*. Paris: Bachelier.
Chasles, Michel. 1828. Géométrie de Situation. Mémoire sur les propriétés des systèmes de sections coniques, situées dans un même plan. *Annales de Mathématiques Pures et Appliquées* 18: 277–301.
———. 1837. *Aperçu Historique sur l'Origine et le Développement des Méthodes en Géométrie Particulièrement de Celles Qui se Rapportent à la Géométrie Moderne*. Paris: Gauthier-Villars.
———. 1863. Rapport sur les travaux mathématiques de M. O. Terquem. *Nouvelles Annales de Mathématiques* 2: 241–251.
Chemla, Karine. 1989. The Background to Gergonne's Treatment of Duality: Spherical Trigonometry in the Late 18th Century. In *The History of Modern Mathematics: Ideas and Their Reception*, ed. John McCleary and David Rowe, vol. 1, 331–359. Boston: Harcourt Brace Jovanovich.
Chemla, Karine, and Serge Pahaut. 1988. Préhistoire de la dualité: explorations algébriques en trigonométrie sphérique 1753–1825. In *Sciences à l'Époque de la Révolution Française*, ed. R. Rashed, 148–200. Paris: A. Blanchard.
Christen-Lécuyer, Carole, and François Vatin, eds. 2009. *Charles Dupin. 1784–1873: Ingénieur, Savant, Économiste, Pédagogue et Parlementaire du Premier au Second Empire*. Rennes: Presses Universitaires de Rennes.
Clairaut, Alexis Claude. 1830. *Élémens de Géométrie*. 5th ed. Paris: Bachelier.
Clebsch, Alfred. 1872. *Zum Gedächtnis an Julius Plücker*. Göttingen: Dieterichsche Buchhandlung.
Coolidge, Julian Lowell. 1934. The rise and fall of projective geometry. *The American Mathematical Monthly* 41 (4): 217–228.
Cournot, Auguste. 1827. 267. Transformation et division des figures sphèriques, au moyen de constructions graphiques; par M. Steiner.. Journ. der Mathem., de Crelle; t. 2, p. 45. *Bulletin des Sciences Mathématiques, Astronomiques, Physiques et Chimiques* 8: 298–303.
———. 2010. *Écrits de Jeunesse et Pièces Diverses*. Vol. 1. Paris: Presses Universitaires de Franche-Comté.
Cremona, Luigi. 1873. *Elementi di Geometria Projettiva*. Rome: Paravia e Comp.
———. 1885. *Elements of Projective Geometry*. Trans. C. Leudesdorf. Oxford: Clarendon Press.
D. 1829. Géométrie. Manuel de Géométrie; par M. Terquem. *Bulletin des Sciences Mathématiques, Astronomiques, Physiques et Chimiques* 11: 1.
D'Enfert, Renaud, and Virginie Fonteneau. 2011. *Espaces de l'Enseignement Scientifique et Technique: Acteurs, Savoirs, Institutions, XVIIe-XXe Siècles*. Paris: Hermann.
de LaCaille, Nicolas-Louis, and Jean-Baptiste Labey. 1811. *Leçons Élémentaires de Mathématiques*. 5th ed. Paris: Courcier.
Delcourt, Jean. 2011. Annales de Gergonne et Nouvelles Annales de la géométrie élémentaire à la géométrie supérieure. *Archive for the History of the Exact Sciences* 65: 1–14.
Despeaux, Sloan Evans. 2008. Mathematics sent across the channel and the Atlantic: British Mathematical contributions to European and American Scientific Journals, 1835–1900. *Annals of Science* 65 (1): 73–99.
Develey, Emmanuel. 1812. *Élémens de Géométrie*. Paris: Vve Courcier.

Dhombres, Jean. 1985. French Mathematical Textbooks from Bezout to Cauchy. *Historia Scientiarum: International Journal of the History of Science Society of Japan* 98: 91–137.

Dhombres, Jean, and Nicole Dhombres. 1989. *Naissance d'un Nouveau Pouvoir: Sciences et Savants en France, 1794–1824*. Paris: Payot.

Didiez, N.J. 1828. *Cours Complet de Géométrie*. Paris: Bachelier.

dos Santos, Cleber Haubrichs. 2015. *Étienne Bobillier. 1798–1840: Parcours Mathématique, Enseignant et Professionel*. Ph.D. thesis, Université de Lorraine and Universidade Federal do Rio de Janeiro, Nancy and Rio de Janeiro.

Dupin, Charles. 1813. *Développements de Géométrie, avec des Applications à la stabilité des Vaisseaux, aux Déblais et Remblais, au Défilement, à l'Optique, etc.; ouvrage approuvé par l'Institut de France, pour Faire Suite à la Géométrie Descriptive et à la Géométrie Analytique de M. Monge*. Paris: Vve Courcier.

———. 1822. *Applications de Géométrie et de Méchanique, à la Marine, aux Ponts et Chaussées, etc., Pour Fiare Suite aux Développements de Géométrie*. Paris: Bachelier.

Ehrhardt, Caroline. 2010. A social history of the 'Galois Affair' at the Paris Academy of Sciences, 1831. *Science in Context* 23: 91–119.

Fano, Gino. 1907. Gegensatz von synthetischer und analytischer Geometrie in seiner historischen Entwicklung im 19. Jahrhundert. *Encyklopädie der Mathematischen Wissenschaften* 3 (4): 221–288.

Flament, Dominique, ed. 1997. *Le Nombre, une Hydre à n Visages; etre Nombres Complexes et Vecteurs*. Paris: Éditions de la Maison des Sciences de L'homme.

Francoeur, Louis-Benjamin. 1828a. *Cours Complet de Mathématiques Pures*. Paris: Bachelier.

———. 1828b. Livres Français. Sciences Physiques. 297. Géométrie appliquée à l'industrie, à l'usage des artistes et des ouriers; par E.-L. Bergery. *Revue Encyclopédique Numéro de Mars* 37: 753–755.

Friedelmeyer, Jean-Pierre. 2011. In *L'impulsion Originelle de PONCELET dans l'Invention de la Géométrie*, ed. Bioesmat-Martagnon, vol. 2011, 55–158.

———. 2016. Quelques jalons pour l'histoire des transformations au 19e siècle, les contributions de Poncelet et Möbius. In *Eléments d'une Biographie de l'Espace Géométrique*, ed. Lise Bioesmat-Martagon, 15–68. Nancy: Presses Universitaires de Nancy.

Garnier, Jean Guillaume. 1808. *Élémens de Géométrie Analytique*. Paris: Courcier.

———. 1812. Recherche de la distance entre les centres des cercles inscrit et circonscrit à un même triangle. *Annales de Mathématiques Pures et Appliquées* 3: 346–347.

———. 1813. *Géométrie Analytique, ou Application de l'Algèbre à la Géométrie*. Paris: Vve Courcier.

Gaultier, Louis. 1813. Mémoire sur les moyens généraux de construire graphiquement les cercles déterminés par trois conditions, et les sphères déterminés par quatre conditions. *Journal de l'Ecole Polytechnique* 16: 124–214.

Gergonne, Joseph-Diez. 1810. Prospectus. *Annales de Mathématiques Pures et Appliquées*, i–v.

———. 1813. Géométrie analitique. Théorie analitique des pôles des lignes et des surfaces du second ordre. *Annales de Mathématiques Pures et Appliquées* 3: 293–302.

———. 1814. Géométrie transcendante. Démonstration des principaux théorèmes de M. Dupin sur la courbure des surfaces. *Annales de Mathématiques Pures et Appliquées* 4: 368–378.

———. 1823. Géométrie élémentaire. Sur la construction du cercle tangent à trois cercles donnés. *Annales de Mathématiques Pures et Appliquées* 13: 193–200.

———. 1826a. Livres Nouveaux. *Annales de Mathématiques Pures et Appliquées* 16 (8). backmatter.

———. 1826b. Philosophie Mathématique. Considérations philosophiques sur les élémens de la science de l'etendue. *Annales de Mathéma-tiques Pures et Appliquées* 16: 209–231.

Gérini, Christian. 2010. Du journal de L'Ecole polytechnique au journal de Liouville: les annales de mathématiques pures et appliquées de Gergonne et les Polytechniciens. *Bulletin de la Sabix* 45: 47–55.

Gérini, Christian, and Norbert Verdier. 2007. Les Annales de Gergonne (1810-1832) et le Journal de Liouville (1836-1874): une mine de textes numérisés à exploiter dans notre enseignement. *Repères IREM* 67: 55–68.

Gispert, Hélène. 2014. Mathematics education in France: 1800–1980. In *Handbook on the History of Mathematics Education*, ed. Alexander Karp and Gert Schubring, 229–240. New York: Springer.

Grattan-Guinness, Ivor. 1984. Work for the workers: Advances in engineering mechanics and instruction in France, 1800–1830. *Annals of Science* 41: 1–33.

Hachette, Jean-Nicholas-Pierre, and Gaspard Monge. 1813. *Traité des Surfaces du Second Degré*. Paris: Klostermann Fils.

Housel, Charles Pierre. 1865. *Introduction à la Géométrie Supérieure*. Paris: Gauthier-Villars.

Lacroix, Sylvestre François. 1799. *Élémens de Géométrie*. Paris: Duprat.

———. 1803a. *Élémens de Géométrie*. 3rd ed. Paris: Courcier.

———. 1803b. *Traité Élémentaire de Trigonométrie Rectiligne et Sphérique, et d'Application de l'Algèbre à la Géométrie*. 3rd ed. Paris: Courcier.

Legendre, Adrien-Marie. 1800. *Éléments de Géométrie*. 3rd ed. Paris: Firmin Didot.

———. 1812. *Éléments de Géométrie*. 9th ed. Paris: Firmin Didot.

———. 1832. *Éléments de Géométrie*. 14th ed. Brussels: H. Remy.

Lombard, Philippe. 2011. On appellera espace projectif. In *Bioesmat-Martagnon*, 13–54.

Loria, Gino. 1887. *Il Passato ed il Presente Delle Principali Teorie Geometriche*. Torino: Carlo Clausen.

Monge, Gaspard. 1798. *Géométrie Descriptive*. Paris: Baudouin.

———. 1807. 1795. *Application de l'Analyse à la Géométrie à l'usage de l'École Impériale Polytechnique*. Paris: Bernard.

———. 1814. Du centre de similitude de deux courbes semblables. *Correspondance sur l'École Polytechnique* 3 (1): 4–5.

Moussard, Guillaume. 2015. *Les Notions de Problèmes et de Méthodes dans les Ouvrages D'enseignement de la Géométrie en France. 1795–1891*. Ph.D. thesis, Université Nantes Angers Le Mans, Nantes.

Mulcahy, John. 1862. *Principles of Modern Geometry*. Dublin: Hodges, Smith & Co.

Mutel, Auguste. 1831. *Cours de Géométrie et de Trigonométrie*. Paris: Mme Ve Bernard.

Nabonnand, Philippe. 2011. Deux droites coplanaires sont sécantes. In *Bioesmat-Martagnon*, vol. 2011, 159–195.

———. 2015. L'étude des propriétés projectives des figures par Poncelet: une modernité explicitement ancrée dans la tradition. In *Sciences Mathématiques 1750–1850: Continuités et ruptures*, ed. Alexandre Guilbaud and Christian Gilain, 381–402. Paris: CNRS Éditions.

———. 2016. Utiliser des éléments imaginaires en géométrie: Carnot, Poncelet, von Staudt et Chasles. In *Éléments d'un Biographie de l'Espace Géométrique*, ed. Lise Bioesmat-Martagon, 69–106. Nancy: PUN-Édulor.

Olivier, G.F. 1835. *Géométrie Usuelle*. 3rd ed. Paris: Maire-Nyon.

Otero, Mario H. 1997. Joseph-Diez Gergonne (1771–1859). *Histoire et Philosophie des Sciences, Sciences et Techniques en Perspective* 37: 1–260.

Peyrard, Fraçois. 1804. *Les Élémens de Géométrie d'Euclide*. Paris: F. Louis.

Plücker, Julius. 1826. Géométrie analytique. Recherche graphique du cercle osculateur, pour les lignes du second ordre. *Annales de Mathématiques Pures et Appliquées* 17: 69–72.

———. 1827. Géométrie analytique. Mémoire sur les contacts et sur les intersections des cercles. *Annales de Mathématiques Pures et Appliquées* 18: 29–47.

Poncelet, Jean Victor. 1817. Philosophie mathématique. Réflexions sur l'usage de l'analise algébrique dans la géométrie; suivies de la solution de quelques problèmes dépendant de la géométrie de la règle. *Annales de Mathématiques Pures et Appliquées* 8: 141–155.

———. 1822. Traité des propriétés projectives des figures. In *Ouvrage utile a Ceux qui s'Occupent des Applications de la Géométrie Descriptive et D'opérations Géométriques Sur le Terrain*. Paris: Bachelier.

Poullet-Delisle, Antoine Charles Marcelin. 1809. *Application de L'Algèbre à la Géométrie*. Paris: Courcier.

Reye, Theodore. 1866. *Geometrie der Lage*. Hannover: C. Rümpler.

Rollet, Laurent, and Philippe Nabonnand. 2013. Un journal pour les mathématiques spéciales: les Nouvelles annales de mathématiques. 1842–1927. *Bulletin de l'Union des Professeurs de Spéciales–Mathématiques et Sciences Physiques* 86: 5–18.

Rowe, David. 1997. In search of Steiner's Ghosts: Imaginary elements in nineteenth century geometry. In *Le Nombre, une Hydre à n Visages; etre Nombres Complexes et Vecteurs*, ed. Dominique Flament, 193–208. Paris: Éditions de la Maison des sciences de l'homme.

Schoenflies, Arthur. 1909. Projektive Geometrie. *Encyklopädie der Mathematischen Wissenschaften mit Einschluss Ihrer Anwendungen* 3 (5): 389–480.

Schubring, Gert. 1987. On the methodology of analysing historical textbooks: Lacroix as textbook author. *For the Learning of Mathematics* 7 (3): 41–51. ("Errata", ibid., 1988, *8: 2*, 51).

———. 2005. *Conflicts between Generalization, Rigor, and Intuition: Number concepts underlying the Development of Analysis in 17-19th Century France and Germany*. New York: Springer.

Schwab, Jacques. 1813. *Élémens de Géométrie*. Nancy: Hissette.

Servois, François-Joseph. 1803. *Solutions peu Connues de Différens Problèmes de Géométrie Pratique*. Paris: Bachelier.

———. 1810. Solution du premier des deux problèmes proposés à la page 259 de ce volume., et du problème proposé à la page 126 du même volume. *Annales de Mathématiques Pures et Appliquées* 1: 337–341.

Smith, David Eugene. 1906. *History of Modern Mathematics*. 4th ed. New York: John Wiley and Sons.

Steiner, Jakob. 1826. Einige geometrische Betrachtungen. *Journal für die Reine und Angewandte Mathematik* 1: 161–182.

Steiner, Jakob, and Joseph-Diez Gergonne. 1827. Géométrie pure. Théorie générale des contacts et des intersections des cercles. *Annales de Mathématiques pures et Appliquéevs* 17: 285–315.

Taton, René. 1949. La préhistoire de la «géométrie moderne». *Revue d'Histoire des Sciences et de leurs Applications* 2 (3): 197–224.

Terquem, Olry. 1829. *Manuel de Géométrie, ou Exposition Élémentaire des Principes de Cette Science*. Paris: Roret.

———. 1859. Sur diverses géométries. *Nouvelles Annales de Mathématiques* 18: 445–446.

Vatin, François, ed. 2007. *Morale Industrielle et Calcul Économique dans le Premier XIXe siècle: L'économie Industrielle de Claude-Lucien Bergery. (1787–1863)*. Paris: L'Harmattan.

Verdier, Norbert. 2009. *Le Journal de Liouville et la presse de son temps: une entreprise d'édition et de circulation des mathématiques au XIXe siécle (1824–1885)*. Ph.D. thesis, Université Paris 11, Paris.

Vincent, Alexandre. 1826. *Cours de Géométrie Élémentaire*. Paris: Bachelier.

———. 1832. *Cours de Géométrie Élémentaire*. 2nd ed. Paris: Bachelier.

———. 1834. *Cours de Géométrie Élémentaire*. 3rd ed. Paris: Bachelier.

Appendix Chronological table of geometry textbooks consulted

Date	Author	Title	Publisher	Place
1798	Gaspard Monge	Géométrie descriptive	Baudouin	Paris
1799	Silvestre-François Lacroix	Élémens de géométrie	Duprat	Paris
1800 (6th edition)	Charles Bossut	Cours de mathématiques	Firmin Didot	Paris
1800 (3rd edition)	Adrien-Marie Legendre	Éléments de géométrie	Firmin Didot	Paris
1802 (2nd edition)	Silvestre-François Lacroix	Essais de géométrie sur les plans et les surfaces courbes: Élémens de Géométrie descriptive	Duprat	Paris
1803 (3rd edition)	Silvestre-François Lacroix	Élémens de géométrie	Courcier	Paris
1803 (3rd edition)	Sylvestre-François Lacroix	Traité élémentaire de trigonométrie rectiligne et sphérique, et d'application de l'algèbre à la géométrie	Courcier	Paris
1804	François Servois	Solutions peu connues de différens problèmes de géométrie-pratique	Bachelier	Paris
1806	Christian Kramp	Élémens de géométrie	Hansen	Cologne
1807 (4th edition)	Silvestre-François Lacroix	Traité élémentaire de trigonométrie rectiligne et sphérique, et d'application de l'algèbre à la géométrie.	Courcier	Paris
1809 (4th edition)	Gaspard Monge	Application de l'Analyse à la Géométrie à l'usage de l'École Impériale Polytechnique	Vve Bernard	Paris
1809	Antoine Charles Marcelin Poullet-Deslile	Application de l'algèbre à la géométrie	Courcier	Paris
1810	Jean-Guillaume Garnier	Réciproques de la géométrie, suivies d'un recueil de théorèmes et de problèmes	Courcier	Paris
1810 (2nd edition)	Jean-Louis Boucharlat	Théorie des courbes et des surfaces du second ordre, précédée des principes fondamentaux de la géométrie analytique	Vve Courcier	Paris

(continued)

Appendix (continued)

Date	Author	Title	Publisher	Place
1810 (4th edition)	Jean-Baptiste Biot	Essai de géométrie analytique, appliqué aux courbes et aux surfaces du second ordre	J. Klostermann fils	Paris
1811 (5th edition)	Nicolas-Louis de LaCaille	Leçons élémentaires de mathématiques	Courcier	Paris
1811 (9th edition)	Silvestre-François Lacroix	Élémens de géométrie	Vve Courcier	Paris
1811	Claude-Jacques Toussaint	Traité de géométrie et d'architecture théorique et pratique, simplifié	Hocquet et Compe	Paris
1812 (2nd edition)	Louis Bertrand	Élémens de géométrie	J. J. Paschoud	Paris
1812	Emanuel Develey	Élémens de géométrie	Vve Courcier	Paris
1812 (4th edition)	Silvestre-François Lacroix	Essais de géométrie sur les plans et les surfaces courbes: Élémens de Géométrie descriptive	Vve Courcier	Paris
1812 (9th edition)	Adrien-Marie Legendre	Éléments de géométrie	Firmin Didot	Paris
1813	Charles Dupin	Développements de géométrie	Vve Courcier	Paris
1813 (6th edition)	Silvestre-François Lacroix	Traité élémentaire de trigonométrie rectiligne et sphérique, et d'application de l'algèbre à la géométrie	Vve Courcier	Paris
1813	Jacques Schwab	Élémens de géométrie	Hissette	Nancy
1813 (5th edition)	Jean-Baptiste Biot	Essai de géométrie analytique, appliqué aux courbes et aux surfaces du second ordre	J. Klostermann fils	Paris
1813	Jean-Guillaume Garnier	Géométrie analytique, ou Application de l'algèbre à la géométrie	Vve Courcier	Paris
1815	J. de Stainville	Mélanges d'analyse algébrique et de géométrie	Vve Courcier	Paris
1816 (2nd edition)	Emanuel Develey	Élémens de géométrie	Vve Courcier	Paris
1817	Jean-Nicholas-Pierre Hachette	Éléments de géométrie à trois dimensions. Partie synthétique et partie algébrique	Vve Courcier	Paris
1817	Charles Michel Potier	Traité de géométrie descriptive	Firmin Didot	Paris

(continued)

Appendix (continued)

Date	Author	Title	Publisher	Place
1818	Gabriel Lamé	Examen des différentes méthodes employées pour résoudre les problèmes de géométrie	Vve Courcier	Paris
1819 (11th edition)	Silvestre-François Lacroix	Élémens de géométrie	Vve Courcier	Paris
1819	Antoine-André-Louis Reynaud	Traité d'application de l'algèbre à la géométrie, et de trigonométrie	Vve Courcier	Paris
1819	Paul-Marie-Gabriel Treuil	Essais de mathématiques, contenant quelques détails sur l'arithmétique, l'algèbre, la géométrie et la statique	Vve Courcier	Paris
1819	Louis-Léger Vallée	Traité de la géométrie descriptive	Vve Courcier	Paris
1821	L. J. George	Essai de géométrie pratique, destiné aux instituteurs primaires aux élèves des collèges	Beaucolin	Neufchateau
1821	Luis-Léger Vallée	Traité de la science du dessin, contenant la théorie générale des ombres, la perspective linéaire, la théorie générale des images d'optique et la perspective aérienne appliquée au lavis, pour faire suite à la Géométrie descriptive	Mme Vve Courcier	Paris
1822 (5th edition)	Silvestre-François Lacroix	Essais de géométrie sur les plans et les surfaces courbes: Élémens de Géométrie descriptive	Bachelier	Paris
1822	Jean-Nicolas Noël	Mélanges de mathématiques, ou Application de l'algèbre à la géométrie élémentaire	C. Lamort	Metz
1822	Charles Dupin	Applications de géométrie et de mécanique à la marine, aux ponts-et-chaussées, etc.	Bachelier	Paris
1823	Alexandre Denuelle	Traité simple et concis de géométrie pratique (2nd edition)	C. L. F. Panckoucke	Paris
1823	Joseph Adhémar	Cours de géométrie descriptive	Chaignieau fils ainé	Paris
1824	A. Person de Teyssèdre	Notions élémentaires d'arithmétique, de géométrie, de mécanique, de physique, de dessin linéaire, perspective et architecture	Fain	Paris
1825	Pierre Louis Marie Bourdon	Application de l'algèbre à la géométrie	Bachelier	Paris
1825	Claude-Lucien Bergery	Géométrie appliquée à l'industrie	Lamort	Metz

(continued)

Appendix (continued)

Date	Author	Title	Publisher	Place
1825	P. A. B. Dupont	Éléments de géométrie théorique et pratique	Boucher	Paris
1825	Charles Dupin	Géométrie et mécanique des arts et métiers et des beaux-arts	Bachelier	Paris
1826 (7th edition)	Jean-Baptiste Biot	Essai de géométrie analytique, appliqué aux courbes et aux surfaces du second ordre	Bachelier	Paris
1826	Claude-Lucien Bergery	Géométrie des sciences industrielles Seconde partie. Géométrie des courbes appliquée à l'industrie	Lamort	Metz
1826	Alexandre Vincent	Cours de géométrie élémentaire	Bachelier	Paris
1826	Nicolas Fourneau	Essais pratiques de géométrie	Firmin Didot	Paris
1826	Pierre Desnanot	Pratique du toisé géométrique, ou Géométrie pratique	Thibaud-Landriot	Clermont-Ferrand
1827	Lancelot	Dessin linéaire et géométrie pratique	Boniez-Lambert	Châlons
1827	A. Person de Teyssèdre	Géométrie des artistes et ouvriers	Decourchant	Paris
1827	Guillaume Henri Dufour	Géométrie perspective	Bachelier	Paris
1827	Louis Gaultier	Notions de géométrie pratique (2nd edition)	L. Colas	Paris
1827 (8th edition)	Sylvestre-François Lacroix	Traité élémentaire de trigonométrie rectiligne et sphérique, et d'application de l'algèbre à la géométrie.	Bachelier	Paris
1827	Louis-Etienne Lefébure de Fourcy	Leçons de géométrie analytique	Bachelier	Paris
1827	A. Lefevre	Applications de la géométrie à la mesure des lignes inaccessibles et des surfaces planes	Bachelier	Paris
1827 (5th edition)	Gaspard Monge (Barnabé Brisson)	Géométrie descriptive	Bachelier	Paris
1828	N. J. Didiez	Cours complet de géométrie	Bachelier	Paris
1828	Charles Dupin	Géométrie et mécanique des arts et métiers et des beaux-arts (2nd edition)	Bachelier	Paris
1828	E. Duchesne	Éléments de géométrie descriptive, à l'usage des élèves qui se destinent à l'École Polytechnique, à l'École militaire, à l'École de marine	H. Balzac	Paris
1828	Gabriel Gascheau	Géométrie descriptive	Bachelier	Paris

(continued)

Appendix (continued)

Date	Author	Title	Publisher	Place
1828 (2nd edition)	Lorenzo Mascheroni (trans. A. M. Carette)	Géométrie du compas	Bachelier	Paris
1828	Claude-Lucien Bergery	Géométrie appliquée à l'industrie (2nd edition)	Lamort	Metz
1828	L. J. George	Géométrie pratique à l'usage des artistes et des ouvriers	C. J. Hissette	Nancy
1828	Émile Martin	Géométrie de l'ouvrier, ou Application de la règle, de l'équerre et du compas à la solution des problèmes de la géométrie	Audot	Paris
1829	Charles Mareschal-Duplessis	La Géométrie des gens due monde	Eberhart	Paris
1829 (2nd edition)	E. Duchesne	Éléments de géométrie descriptive, à l'usage des élèves qui se destinent à l'École Polytechnique, à l'École militaire, à l'École de marine	H. Balzac	Paris
1829 (3rd edition)	Enrico Giamboni (trans. D. Roux)	Éléments d'algèbre, d'arithmétique et de géométrie, où l'arithmétique et la géométrie se déduisent des premières notions de l'algèbre	Bachelier	Paris
1829	Amand-Denis Vergnaud	Manuel de perspective du dessinateur et du peintre (third edition)	Roret	Paris
1829	Olry Terquem	Manuel de géométrie, ou exposition élémentaire des principes de cette science	Roret	Paris
1829	Enrico Giamboni	Éléments d'algèbre, d'arithmétique et de géométrie, où l'arithmétique et la géométrie se déduisent des premières notions de l'algèbre (translated from edition)	Bachelier	Paris
1830 (1741)	Alexis-Claude Clairaut	Élémens de géométrie	Bachelier	Paris
1830	Louis Gaultier	Notions de géométrie pratique (2nd edition)	J. Renouard	Paris
1830 (14th edition)	Silvestre François Lacroix	Élémens de géométrie	Bachelier	Paris
1830	H. Vernier	Géométrie élémentaire à l'usage des classes d'humanités et des écoles primaires	L. Hachette	Paris
1830	Hippolyte Véron Vernier	Géométrie élémentaire à l'usage des classes d'humanités et des écoles primaires	A. Felin	Paris

(continued)

Appendix (continued)

Date	Author	Title	Publisher	Place
1831	Auguste Mutel	Cours de géométrie et de trigonométrie	Vve Bernard	Paris
1831	Claude-Lucien Bergery	Géométrie des écoles primaires	P. Wittersheim	Metz
1831	Mathieu Bransiet	Abrégé de géométrie pratique appliquée au dessin linéaire	Moronval	Paris
1832	A. Delhorbe	Nouveau Traité de géométrie pratique	Guyot-Roblet	Rheims
1832	François-Cheri Duhousset	Application de la géométrie à la topographie	Migneret	Paris
1832 (14th edition)	Adrien-Marie Legendre	Éléments de géométrie	H. Remy	Brussels
1832	Alexandre Meissas	Cours de géométrie	A. Pihan Delaforest	Paris
1833	Antoine-André-Louis Reynaud	Théorèmes et problèmes de géométrie	Bachelier	Paris
1833	Alphonse-Louise-Bernard Boubée Lespin	Traité de géométrie et d'arpentage	Lecointe et Pougin	Paris
1835 (3rd edition)	G. F. Olivier	Géométrie usuelle	Maire-Nyon	Paris

Chapter 5
The Impact of Teaching Mathematics Upon the Development of Mathematical Practices

Gert Schubring

Abstract This chapter discusses interfaces between the development of mathematics and the teaching of mathematics. Contrary to traditional convictions of teaching as being restricted to a receptive and passive role, productive interactions between the two poles are analysed here. Four cases even for an impact of teaching upon mathematical practices will be presented and discussed, featuring the issue of elements and elementarisation, the institutional impact of teacher education on research in pure mathematics, and the dissemination of set theory and of non-Euclidean geometry by German school textbooks in the second half of the nineteenth century.

Keywords D'Alembert · Destutt de Tracy · Elements · Elementarisation · Friedrich Meyer, Hermann Wagner · Interfaces · Non-Euclidean geometry · Richard Dedekind · Set theory · Teacher education

1 Introduction: Issues of Methodology

Traditionally, in mathematics and historiography, the teaching of mathematics has been seen as having no influence on mathematical practices and their development. The contents of teaching are seen as a certain kind of projection of academic mathematics, as a certain sedimentation. Therefore, the relation between the development of mathematical practices and the teaching of mathematics is often conceived of as unilateral, without an impact of teaching upon research. This chapter undertakes it to show that there are productive interactions between the two poles. Four cases for an impact of teaching upon mathematical practices will be presented and discussed. While the first one will discuss the importance of the notion of element and elementarisation in the interface between mathematical development and teaching, and the

G. Schubring (✉)
Universidade Federal do Rio de Janeiro (UFRJ), Rio de Janeiro, Brazil

Universität Bielefeld, Bielefeld, Germany
e-mail: gert.schubring@uni-bielefeld.de

G. Schubring (ed.), *Interfaces between Mathematical Practices and Mathematical Education*, International Studies in the History of Mathematics and its Teaching,
https://doi.org/10.1007/978-3-030-01617-3_5

second will discuss the impact of teacher education for research profiles, the third and the fourth will discuss conceptual developments of nineteenth century mathematics.

A paradigmatic case for the traditional position is the French mathematics educator Yves Chevallard who has made it the foundation of a theoretical generalisation, elaborated as *transposition didactique* and widely influential in mathematics education. The conception of the didactic transposition proposes to examine how academic knowledge of mathematics ("savoir savant") becomes school mathematical knowledge. For this, Chevallard distinguished three types of knowledge:

- "Objet de savoir"—object of knowledge, i.e. the knowledge achieved by mathematics.
- "Objet d'enseigner"—subject to be taught: The academic knowledge becomes teachable knowledge by the efforts of mathematics educators (their community being called "noosphère").
- "Objet d'enseignement"—teaching subject: The subject to be taught as adapted by teachers (Chevallard 1985, 39).

Analysing this conception, it becomes clear that the transposition notion offered conceives of a unilateral process: it has as its starting point, a pole designed as advanced, the academic or university knowledge and as its final point another pole inferior to it, occurring in schools and involving the teacher in the classroom.

Willem Kuyk—the author of "Complementarity in mathematics" (1977)—, however, had denounced this traditional view in 1978, in stating: "Mathematics is not a stalactite hanging over a stalagmite"; Kuyk thus denied the view that mathematics education grows only by receiving some drops from above, from the supreme instance (Schubring 2001, p. 297).

A historiographical endeavour where one might expect a reflection about the interfaces between mathematical research and the teaching of mathematics is the monumental work *Writing the History of Mathematics*, edited by Joseph Dauben and Christoph Scriba in 2002, where the historiography of mathematics is analysed in a most comprehensive chronological and geographical manner. Yet, given the fact that historiography of mathematics had largely been written by mathematicians, historiography followed essentially the preoccupations of mathematicians "with respect to chronology and where questions about priorities and the actual sequence of internal mathematical developments are concerned", given their primary interest "in the history of concepts and methods" (Dauben and Christoph 2002, p. xxiv).

Thus, although the editors asked, "are there any general historiographic principles that emerge from these studies, ones that seem to transcend time and national boundaries?" (ibid., p. xxiii), the study does not go beyond what the respective analysed authors had elaborated from their traditional viewpoint. Being descriptive, the volume documents that historiography was practised until very recently by mathematicians—with the notable exception of France, where philosophers and *épistémologues* took the lead in the twentieth century—and that their focus was on the internal history of ideas. The professionalisation of historiography dates from recent times. While the section "History of Mathematics and Mathematics Education" in the *Postscriptum* might have addressed new functions of mathematics education, it remains restricted, however, to the use of history in teaching mathematics. The following section "History of Mathematics: Recent Trends" does not address interfaces between research and teaching (ibid., pp. 335 ff.).

While this volume documents that mathematics historiography is still strongly marked by the opposition between "internal" and "external" approaches, a new German *Handbuch Wissenschaftsgeschichte* of 2017 declares this dispute as overcome and is open to much broader conceptual approaches, understanding science as just one form of knowledge—history of science being hence a part of *Wissensgeschichte*, the history of knowledge (Sommer et al. 2017, p. 3). In fact, this handbook realised an ambitious endeavour to reflect the methodology of history of science research; it presents, in particular, systematic chapters on recent research approaches. Pertinent for research on our issue of interfaces is the chapter on cultural sciences and science history (Brandt 2017). Likewise, the series of chapters on places of knowledge production is novel.

2 Examples for Introducing the Interface Approach

To give the first piece of evidence for the productive role of teaching: as is well known, Dedekind emphasised in the preface of his book *Stetigkeit und irrationale Zahlen* (1872), which became decisive for establishing a rigorous concept of real numbers, that it was his experience in teaching the infinitesimal calculus for the first time at the *Eidgenössische Technische Hochschule* (ETH) Zürich, in 1858, that made him conscious of the missing fundaments for the number concept (see Fig. 5.1):

> The reflections which form the subject of this little work date from the autumn of 1858. At that time, as a professor at the Swiss Polytechnic in Zurich, I was in a position to lecture the elements of differential calculus for the first time, and felt more sensitive than ever before to the lack of a truly scientific justification of arithmetic. Regarding the concept of a variable quantity approaching a fixed limit, and especially in the proof of the proposition that every quantity which grows steadily, but not beyond all limits, must certainly approach a limit, I resorted to geometrical evidence (Dedekind 1872, p. 1).[1]

To add a second piece of evidence: Belhoste recalled that the project which initiated the collective work of the Bourbaki group in the 1930s was to elaborate an analysis textbook: it was intended to be more modern in particular than the textbook by Édouard Goursat, *Cours d'analyse mathématique*, first published in 1902 and dominant in France since the early twentieth century (Belhoste 1998, p. 300). In beginning this initially restricted task, the group was lead to search for the foundations of

[1] Die Betrachtungen, welche den Gegenstand dieser kleinen Schrift bilden, stammen aus dem Herbst des Jahres 1858. Ich befand mich damals als Professor am eidgenössischen Polytechnikum zu Zürich zum erste Male in der Lage, die Elemente der Differentialrechnung vortragen zu müssen, und fühlte dabei empfindlicher als jemals früher den Mangel einer wirklich wissenschaftlichen Begründung der Arithmetik. Bei dem Begriffe der Annäherung einer veränderlichen Größe an einen festen Grenzwerth und namentlich bei dem Beweise des Satzes, daß jede Größe, welche beständig, aber nicht über alle Grenzen wächst, sich gewiß einem Grenzwerth nähern muß, nahm ich meine Zuflucht zu geometrischen Evidenzen.

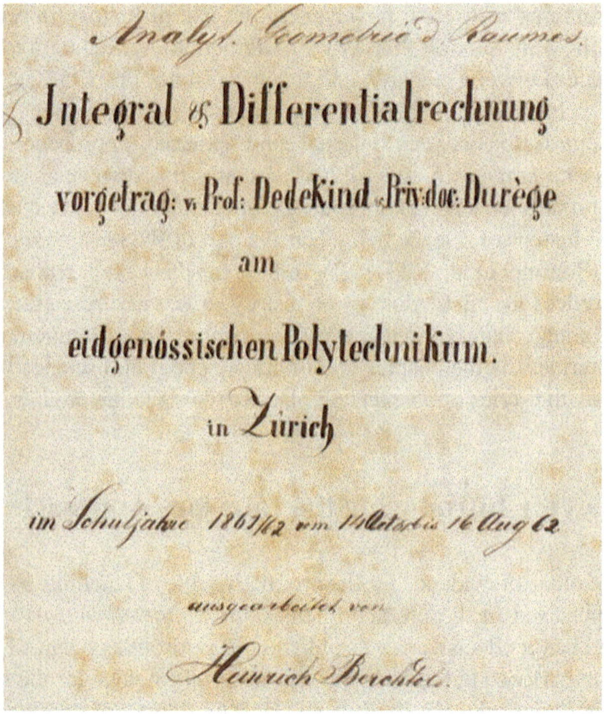

Fig. 5.1 Lecture notes of Dedekind's differential calculus lectures, taken by the student Heinrich Berchtold in the winter term 1861/62 (the next term, due to Dedekind's move to Braunschweig, Heinrich Durège continued the course). ETH-Bibliothek

analysis—so that their focus became to elaborate the textbook on set theory as the first *fascicule*. This turned out to be so complicated and challenging that this volume, *Théorie des ensembles*, took a long time to be ready for publishing. Bourbaki's search for a rigorous presentation of analysis had thus even more profound and comprehensive outcomes than Dedekind's search—as evidenced by the common title of Bourbaki's work: *Éléments de mathématique*. Both historically significant examples reveal us the key function of the notion of *element*.

3 The Notion of Element and of Elementarisation

In fact, the notion of element connects the development of mathematics and the modes of teaching mathematics in a fundamental way. Since Euclid's geometry textbook, the term "elements" expresses the intention to give a systematic presentation of a mathematical theory, constructed from its basic components (see Trouvé 2008, pp. 21 ff.). While thus fixing the state of knowledge of mathematics or of one of its branches for a certain time and period, a textbook represents a stage in the development of mathematics. At the same time, such a textbook provides the material for teaching mathematics.

Remarkably, it was in France that the notion of elements and its role for textbooks and in particular for elementarising knowledge was reflected most explicitly. The reflections began practically right with Modern Times with criticism of how Euclid elementarised mathematics: Pierre de la Ramée or Petrus Ramus (1515–1572) refuted Euclid's *Elements* as the model of a rigorous and methodical presentation of mathematics. Ramus did not just criticise particular propositions or the exactitude or rigour of certain statements, but much more fundamentally their methodology. In Ramus's view, the Elements were not, as traditionally judged, the primordial model for rigorous reasoning and for logical deduction, but rather revealed a lack of a natural, methodical order. Ramus, on the one hand, developed rules for methodical thought, and on the other hand proposed an entirely different order and architecture for mathematics: that it should begin with the general—the general being, in Ramus's view, not geometry, but arithmetic. In addition, arithmetic and geometry should be treated first separately and then combined (Ramus 1569).

Ramus's approach was continued and perfected by a new type of textbook that realised his methodological conceptions: by Antoine Arnauld (1612–1694), a Jansenist philosopher and theologian. Arnauld dared, for the first time, to challenge Euclid's model by claiming to be able to realise an alternative and better model, the title of his textbook is already emblematic and programmatic: ***Nouveaux** élémens de géométrie* (1667), with the subtitle "contenant Outre un ordre tout nouveau, & de nouvelles demonstrations des propositions les plus communes" expressing this ambition.

The reflections on elementarisation were taken up and deepened by d'Alembert in the Encyclopédie, as an essential part of the Enlightenment programme to make knowledge generally accessible. There is an extensive entry in the Encyclopédie, "élémens des sciences", where d'Alembert published these reflections. He started from a first, rough distinction, calling "elements" the first and original components of a whole:

> On appelle en général *élémens d'un tout*, les parties primitives & originaires dont on peut supposer que ce tout est formé. (d'Alembert 1755, col. 491, l)

According to him, it would be easy to identify these original parts, which serve as basis:

> il est facile de distinguer les propositions ou vérités générales qui servent de base aux autres, & dans lesquelles celles-ci sont implicitement renfermées. (d'Alembert 1755, col. 491r)

That the other, more developed concepts would be implicitly enclosed in the basic ones reveals d'Alembert's specific conception of elementarisation, since he continued:

> Ces propositions réunies en un corps, formeront, à proprement parler, les élémens de la science, puisque ces *élémens* seront comme un germe qu'il suffiroit de développer pour connoitre les objets de la science fort en détail.[2] (ibid.)

The key term here is "germ". And this biological analogy means that the element is a kind of nucleus, which already contains all possibilities of unfolding, of development. Its unfolding will hence result in a coherent structure. It is in this sense that Bourbaki used to speak of the architecture of mathematics (see Bourbaki 1948). "Elementarising", therefore, means to expose a mathematical theory as structured and built from its elements, understood in this way.

D'Alembert expressed the characteristic optimistic vision of the Enlightenment that, by this elementarisation, knowledge can be universally disseminated and understood:

> Tout ce qui est vrai, surtout dans les sciences de pur raisonnement, a toûjours des principes clairs & sensibles, & par conséquent peut être mis à la portée de tout le monde sans aucune obscurité (d'Alembert 1755, col. 492r).[3]

D'Alembert thus launched the conception of *livres élémentaires*, intended to be the primary preoccupation for education in the first stages of the French Revolution. D'Alembert had called on scientists to elaborate these textbooks, criticising that they so far preferred to strive for their personal fame:

> Uniquement occupés de faire de nouveaux progrès dans l'art, pour s'élever, s'il leur est possible, au-dessus de leurs prédécesseurs ou de leurs contemporains, & plus jaloux de l'admiration que de la reconnoissance publique, ils ne pensent qu'à découvrir & à jouir, & préfèrent la gloire d'augmenter l'édifice au soin d'en éclairer l'entrée (d'Alembert 1755, col. 496r).[4]

This conception of *livres élémentaires* became adopted during the first stages of the French Revolution; one of the first plans for a system of public education, by Talleyrand, postulated:

> Il faut [...] que des livres élémentaires [...] rendent universellement familières toutes les verités (quoted from Schubring 1988, p. 160).[5]

The first projects for a new educational system were in fact based on elaborating livres élémentaires. In 1792, L. F. A. Arbogast proposed a *concours* for composing such textbooks for the disciplines of the primary schools to be created. He emphasised the urgency in order to have the new books before opening the schools:

---

[2] These propositions united in a body will, properly speaking, form the elements of science, since these elements will be like a germ from which it would be sufficient to develop knowledge of the objects of science in great detail.

[3] All that is true, especially in the sciences of pure reasoning, always has clear and sensible principles, and consequently can be made accessible to everyone without any obscurity.

[4] Only occupied with making new progress in their science, in order to rise, if possible, above their predecessors or their contemporaries, and more jealous of admiration than of public recognition, they intend only to discover and enjoy, and prefer the glory of increasing the building of science rather than take care to light its entrance.

[5] It is necessary [...] that *livres élémentaires* [...] turn all truths universally familiar.

> le moyen le plus efficace pour la régénération de l'enseignement, c'est la composition des livres élémentaires. Il étoit de la plus grande urgence [...] de hâter la composition de ces ouvrages (Arbogast 1792, p. 2).

Yet, due to political problems, it took until 18 January of 1794 for the *concours* to be decided by Parliament; within five months, manuscripts for those *livres* were to be submitted for the ten projected teaching subjects in the primary schools. However, the process of composing proved to be much longer. The final evaluation occurred only after one and a half year. And the results were disappointing: for all the ten disciplines, only seven manuscripts were judged to be qualified. Already in October 1794, Joseph Lakanal gave an intermediary evaluation of the *concours* process. In criticising the conception of many submitted manuscripts, he confirmed and elaborated d'Alembert's conception of elementarisation:

> qui avaient confondus généralement deux objets très différents, des *élémentaires* avec des *abregés*. Resserrer, coarcter un long ouvrage, c'est l'abréger; présenter les *premiers germes* et en quelque sorte la *matrice* d'une science, c'est l'élémenter: ainsi, l'abrégé, c'est précisement l'opposé de l'élémentaire (quoted from Schubring 1988, p. 161).[6]

Noteworthy in particular is the opposition between an abridged handbook and a truly elementarised textbook, characterised here not only by the term "germ", but also by "matrix".

An even more revealing result of the experience with this first *concours* for textbooks for public schools proved to be a deepened understanding of the inter-relation between research and teaching. It was the French philosopher Destutt de Tracy (1754-1836), one of the leading *idéologues*—the then influential French group of philosophers—who evaluated in 1801 the project to elaborate the "livres élémentaires", meant to be the basis for this profoundly new type of teaching. Among the various reasons for the few results of this effort, Destutt de Tracy had outlined that composing a textbook frequently leads the author to tasks of research:

> Often, in exposing a fact, one remarks that this requires new observations, and, when examined more thoroughly, it presents itself in a completely different light: on other occasions it is the principles themselves which have to be revised, or, to connect them with each other, there are many gaps to be filled; in a word, it is not only a question of exposing the truth, but of discovering it (Destutt de Tracy 1801, pp. 4–5; my transl., G.S.).[7]

This assessment of the historical experience reveals a decisive pattern for the interface between the development of mathematics and its teaching: upon preparing teaching—either as oral lecture or as written textbook—one will become aware of missing connections in a logical deduction or remark on problems in the

---

[6] who had generally confounded two very different objects, the elementary with abbreviated ones. To constrict, to coarct a long work, is to shorten it; to present the first germs and, in a way, the matrix of a science is to elementarise it: thus, the abridged is exactly the opposite of the elementary.

[7] Souvent, en rendant compte d'un fait, on s'aperçoit qu'il exige de nouvelles observations, et, mieux examiné, il se présente sous un tout autre aspect: d'autres fois, ce sont les principes eux-mêmes qui sont à refaire, ou, pour les lier entre eux, il y a beaucoup de lacunes à remplir; en un mot, il ne s'agit pas seulement d'exposer la vérité, mais de la découvrir.

foundations of the theory so that one is incited to research for providing the needed conceptions.

It is likewise characteristic that, in this context of reflection about the elementarisation of science, the role of the textbook author also became investigated and even credited. While the share of textbook composition in establishing the elements of science was valued, the textbook author was also assessed in his productive contribution to science. A first such crediting was published in 1796, in a review of the second edition of J.A.J. Cousin's calculus textbook: *Leçons de Calcul Différentiel et de Calcul Intégral* (1796). The review was published in *La Décade*, the journal of the *idéologues*. Its anonymous author assumed the novel stance of attributing to a textbook author the rank of "inventor"—a notion in the discourse on science, that designated an innovative scientist since Clairaut and d'Alembert:

> The author of an elementary book attains the rank of an inventor if he can present the elements, first, in the best order, in the most simple and the most clear manner: if he removes from the science all its technical wrapping and if he illustrates after each step the space traversed in such a manner that the student always knows well where he is (quoted from Schubring 1987, p. 43).

And Sylvestre-François Lacroix (1765–1843), the prolific and successful textbook author since the first periods of the French Revolution, was distinguished even by the *Institut*—the new form of the Academy of Sciences since the Revolution—in being attributed a rank equal to an inventor. The distinction had been given in the *Institut's* report on the project presented by Lacroix to publish a treatise on the differential and integral calculus. In fact, he published this treatise as a three volumes textbook from 1797 to 1799. The report explained:

> To present difficult theories with clarity, to connect them with other known theories, to dismantle some of the systematic or erroneous parts which might have obscured them at the time of their emergence, to spread an equal degree of enlightenment and precision over the whole; or, put shortly: to produce a book which is at the same time elementary and up to the mark in science. This is the objective which Citizen Lacroix has taken to himself and which he could not have attained without engaging himself in profound research and by progressing often at the same level as the inventors (quoted from: ibid.).

This assessment, made still in an Enlightenment period, expresses in a paradigmatic manner the programme of elementarisation and the interface between research and teaching.

4 The Impact of Teacher Education

Recent research upon the social history of mathematics confirms the decisive role of teaching for the progress of research practices. In fact, it was the establishing of study courses for mathematics teacher training in higher education which proved to constitute the predominant structural pattern, initiating for the first time within universities the enabling of research activities for the professors performing the lectures and supervisions for this teacher education study course.

Before the French Revolution, mathematics basically could not be studied for obtaining a degree in mathematics. Lectures by mathematics professors in the Arts Faculty had either a propaedeutic character, preparing for professional studies (and degrees) in one of the three higher faculties, or were encyclopaedic, for a broad, non-specialised use. The first time that proper study courses were established for mathematics occurred as a part of Marquis de Pombal's profound university reforms of Coimbra University in Portugal from 1772: not only was the first Mathematics Faculty created then, but likewise a study course, leading to degrees which should assure its graduates the best teaching positions in the country (Silva 1991).

Admittedly, there were only few graduates of these study courses, but one of the first graduates was Francisco de Borja Garçao Stockler (1759–1829) who published important research about the fundaments of analysis from 1794.

The profound reforms of the educational system in Prussia had a much more far-reaching effect from 1806 onward: the Philosophy Faculty became upgraded, providing for the first time proper degrees for professions—namely and noteworthy for the teaching profession. The mathematics teaching profession was included, since mathematics became one of the three major teaching disciplines at the likewise reformed secondary schools, the *Gymnasien*. Within two decades, the profile of the mathematics professors at the Prussian universities changed completely: the formerly encyclopaedic lectures became replaced by specialised high-level lectures and the professors themselves became specialised researchers—in marked contrast to the other German states where the traditional patterns were continued until up to the middle of the nineteenth century (Schubring 1991a).

The *facultas* degree for teaching at secondary schools remained the only degree throughout the entire nineteenth century, which could be obtained by studying mathematics—the same period, which is renowned and famous for the establishment of the new era of rigour by German mathematics! A second degree—the diploma for applying mathematics in other professions and for higher education careers—became established only in 1942. The doctorate as a degree independent of the teaching profession and leading to university careers had been sought for and achieved only by a few students since about the late 1860s (Schubring 1990).

This key role of teacher training for professionalising mathematical research and constituting mathematical communities is even confirmed by a more recent example from Brazil. Upon the establishment of higher education in Brazil in 1810, mathematics could not be studied as a proper study course, but the mathematics lectures functioned as service courses for engineering professions at military academies and polytechnic schools. No universities were founded throughout the nineteenth century, due to the model function of the French higher education structure of *écoles spéciales* (Schubring 1991b). Universities were founded in Brazil only from the 1930s on, due to changed social and cultural conditions. And then, in the first two universities—the *Universidade de São Paulo* (USP) and *Universidade do Distrito Federal* (UDF), resp. the *Universidade do Brasil* (in Rio de Janeiro) —it was the study course for the *magistério*, the teaching profession, within the equivalent of a Philosophy Faculty, which enabled a "take-off" of practising mathematical research (Pereira 2017).

The first university to be founded was the USP, in 1934. Its distinctive new feature was a Faculty, which basically resembled the German Faculties of Philosophy: the FFCL—*Faculdade de Filosofia, Ciências e Letras*—which constituted in fact the kernel of disciplinary development. The founding decree of the USP, of 25 January 1934, in art. 5, § 1, stipulated the introduction of the teaching licence for those trained to become teachers at secondary schools as the "*licença para o magistério secundário*". The degree afforded studies of a scientific discipline at the FFCL and accompanying pedagogical studies at the Institute of Education, attached to the Faculty. It is even more revealing that the statutes projected doctoral studies; for such studies, only students having the *licenciado* diploma were mentioned to be admitted for an additional two years of studies (§ 12 of the decree).[8] Hence, a direct continuation was established: studying for a teaching licence, and possible continuation for a doctorate.

At the UDF, founded in 1935, here, too, there was a new Faculty besides the integration of various former professional schools, like the polytechnic schools, which was at first called *Escola de Ciências*. It had as its principal function the formation of teachers for secondary schools. The § 25 of the founding statutes, of 5 April 1935, attributed the function of providing study courses for the "candidato ao professorado secundário das ciências" in four different courses: for teachers of mathematics, physics, chemistry, and natural sciences. Doctoral studies were not yet instituted (Fávero and de Castro Lopes 2009, pp. 193–194).

5 An Early Teaching of Set Theory in Germany

In 1885, when Georg Cantor was still perfecting his set theory providing new fundaments for mathematics, Friedrich Meyer (1842–1898) —friend of Cantor and mathematics teacher at the Gymnasium in Halle—elaborated a schoolbook on arithmetic and algebra, as reorganised from this basis in set theory.

The fact that the transposition of new knowledge into school knowledge does not necessarily take a path through the scientific community is shown by set theory, which is regarded as the key example of imposing scientific concepts into school teaching: to my own surprise, in my research on the development of school mathematics in the nineteenth century, I encountered a textbook which was not only the first implementation of Cantor's set theory, but which also propagated a radical reconstruction of arithmetic and algebra for schools on the basis of the concepts of set theory, seventy years before the corresponding effect of Bourbaki on school mathematics. It is the book by Friedrich Meyer: *Elemente der Arithmetik und Algebra*, of 1885.

[8] Source: https://www.al.sp.gov.br/repositorio/legislacao/decreto/1934/decreto-6283-25.01.1934.html. I am grateful to Prof. Rogério Monteiro de Siqueira (USP, Sao Paulo) for communicating me these sources.

Meyer was born near Kulm in East-Prussia in 1842, and he completed the Gymnasium in Kulm and studied in Breslau and Halle, but above all at the University of Berlin, mainly with the number theorist Ernst Eduard Kummer. In 1868 he became a teacher of mathematics at the Gymnasium in Halle, where he worked as a highly respected and highly renowned scholar and educator until his early death in 1898. His extensive scholarship was praised in particular, even beyond mathematics and the natural sciences (Hoffmann 1899). In 1894, he received an honorary doctorate from the University of Halle, especially because of his set theory textbook.[9] Cantor himself greatly appreciated this book and recommended it especially to mathematics teachers (Hoffmann 1899).[10] Wilhelm Lorey, known both as a historian of mathematics and of mathematics teaching, he emphasised the importance of this book in his address for the celebration of Cantor's 70th birthday:

> [Meyer] was one of the first to recognize the far-reaching significance of your ideas. In a time when the scientific world was opposed to you, also in our own country, he had already presented the basic concepts of set theory in his textbook destined for schools, and in the foreword he recommended the study of your writings intensely to the teachers of mathematics (Lorey 1915, p. 273).

Meyer's achievement is all the more significant as Cantor's ideas of set theory were not yet completely elaborated in 1885; in their most elaborate form, they were only published in 1895/96 (important parts were accessible since 1883). It turns out that Cantor's ideas were for Meyer, in effect, only a trigger for developing fundamental concepts that had already been developed by mathematics teachers for a long time. In fact, Meyer did not present the concept of a set as something new, but as belonging to a tradition going back to the ancient Greeks; he referred in particular to Nikomachus (about 100 CE).

I have also shown earlier that Johann Schultz (1739–1805) used the set concept for his infinity concept in 1788, and especially for his attempts to prove the 11th postulate on parallel lines. The notion of set ("Menge") was for him well known and used abundantly for conceiving of infinite sets of numbers, and Schultz developed the number concept based on the set concept (Schubring, 1982). Moreover, in the 1810s and 1820s in Germany, when the programme of algebraisation was still in practice and not yet substituted by the return to valuing synthetic geometry, school textbooks existed that constructed arithmetic and the number concept from the basis of the set notion. Two such examples are the arithmetic textbooks by Mathias Metternich (1818), in Hesse, and by Carl Seebold (1821), in Hesse (see Schubring 1991a, b, p. 190).

Meyers's book is also particularly interesting as a dissemination of Cantor's concepts. Walter Purkert was able to show that Cantor was not surprised by the antinomies of set theory because they had been known to him for a long time and because he had assumed that he had excluded inconsistent multiplicities through his

[9] Information from the archives of the University of Halle-Wittenberg.

[10] Yet, in a letter to the Swedish mathematician Ivar Bendixson with whom he was cooperating on set theory, Cantor expressed some doubts regarding the rigour of Meyer's proofs in this schoolbook (Purkert & Ilgauds 1987, p. 132).

definitions of concepts. In letters to Hilbert, Cantor explained that his 1895 definition of a set (summary of certain well-defined objects [...] to a whole) served the function of excluding inconsistent sets, and already in his formulation of 1883:

> "Every Multiplicity which one is able to think of as a One" (Purkert 1986, pp. 18ff.; also Purkert & Ilgauds 1987).

Remarkably, Meyer also adopted this 1883 definition of a set in his textbook, saying: "Im Begriffe der Menge wird vieles zu einem verbunden" (Meyer 1885, p. 1).[11]

The 1885 edition is given as the second edition, but it was not possible to find a printed first edition; probably, it had circulated only as a manuscript among Meyer's colleagues. In fact, Meyer presented his book as serving for cooperation between the mathematics teachers of his Gymnasium and the school's students for repetition of the teaching in the classroom (ibid., p. iii). From various indications, it becomes clear that Meyer had used it in the upper grades (e.g. Meyer 1891, p. 29).

The preface begins with a rather epistemological discussion. The basis of introducing set theory is the notion of "Anzahl",[12] which has according to Meyer the status of a category—in the Kantian sense, like space and time, thus as a given dimension of thinking. By contrast, all other numbers of elementary arithmetic not being an "Anzahl", i.e. not positive integer numbers, are qualified as "inventions", due to the capacity of equations to generate new number types. At the end of the preface, Meyer strongly recommended the study of Cantor's publications, referring in particular to those collected in Volume II (December 1883) of the Swedish journal *Acta Mathematica*.

The first chapter on *Anzahl, Zählen. Zahlzeichen* introduces set as first notion, being presented as likewise primary, like time and space. As its first characteristic, the notion of *Mächtigkeit*—potency—is presented, discussing sets of equal and unequal potency and finite and infinite sets (Meyer 1885, p. 1). Though a finite set admits an ordering, not only an ordered set is defined, but also an "wohlgeordnet", a well-ordered set (ibid., p. 2). The notion of denumerability follows immediately, by determining two well-ordered sets as denumerably related when each element of the one can be related to one of the other. After this, the definition of "Anzahl" is introduced as a general concept or category, comprising well-ordered mutually denumerable sets (ibid., p. 3). Propositions on well-ordered denumerable sets follow; in particular, complete mathematical induction is presented. The number concept is then derived from the concept of *Anzahl* and *Mächtigkeit*. At first, the signs of the first nine "Anzahlen" are explained and, then, how to count the elements of a

---

[11] In the concept of set a Multiplicity is connected to a One.

[12] Actually, the English language has no translation for "Anzahl" that would distinguish it from "number" for "Zahl". Dictionaries only give "number". Joseph W. Dauben, in his publications about the history of Cantor's set theory, uses "numbering". He draw my attention to a paper by W. W. Tait which relates controversies about an adapted English translation of *Anzahl*: "counting number" versus "enumeral" (Tait 2000, p. 275). Another translation of Cantor's works uses to put just "*number (Anzahl)*", or simply "number"—in italics—, as in the translation of Cantor's treatise on *Grundlagen der Mannigfaltigkeitslehre* of 1883, by Uwe Papart, in *The Campaigner*, vol. 9, no.s 1 & 2. I will use here the German term, in italics.

finite set. The introduction of ordinal and cardinal numbers follows (ibid., p. 6). To operate with the numbers, the signs for "greater than" and "less than" are introduced—and, to assure generality not for concrete numbers but only for general signs of numbers, namely for "letter numbers" *a, b, c, ... x, y, z*. The textbook shows to be an axiomatically structured textbook, which is already quite modern. Thus, in operating with numbers, one finds them presented via the axioms of identity, commutativity, associativity, and distributivity (ibid., p. 8 & 22).

The next chapter is devoted to the operations of the first kind, adding and subtracting. Here the domain of "Anzahl", of positive integers, is extended to relative numbers, i.e. to positive and negative integers and to zero, by extending the operations of addition and subtraction already introduced (ibid., pp. 16 ff.). Another chapter introduces multiplication and division as operations of the second kind, for this extended domain. The product is defined here in set theory terms:

> § 27. Sind *a* und *b* Anzahlen, ist ferner $\mathfrak{A}', \mathfrak{A}'', \mathfrak{A}''', \ldots$ eine wohlgeordnete Menge von der Anzahl *b*, während jedes der Elemente *A* selbst eine wohlgeordnete Menge von der Anzahl *a* ist, so entsteht durch Auflösung eines jeden $\mathfrak{A}$ in seine Elemente wiederum eine wohlgeordnete Menge, deren Anzahl das Produkt aus *a* und *b* genannt und durch *ab* [...] bezeichnet wird (ibid., p. 20).[13]

Then, the extensions to rational and to irrational numbers are described before decimal fractions. In another part, series of numbers are introduced to be continuous in order to use them to introduce real numbers. A third part deals with the operations of the third kind: to exponentiate, square root and logarithmise. Further chapters are: imaginary and complex numbers, theory of equations, theory of permutations and combinations, progressions, theorems from number theory, and finally continued fractions—actually quite demanding subjects.

An examination of the impact of Meyer's textbook also produces remarkable results. At first the book seems to have left his colleagues speechless, for in the three major journals—the *Archiv für Mathematics und Physik*, the *Zeitschrift für mathematischen und naturwissenschaftlichen Unterricht* and the *Zeitschrift für Mathematik und Physik*—which constantly published reviews of a large number of mathematics textbooks, no review of Meyer's book appeared. Eventually, however, the book had not only found the recognition of leading specialists, for instance by Max Simon, but also provided ready access to Cantor's set theory for mathematics teachers (Lietzmann 1909, p. 61). Lietzmann, the leading mathematics educator in Germany since the 1910s, was sceptical as to whether set theory could be taught as the fundament for arithmetic teaching, as is clear from his remark:

> Unfortunately, it is impossible to infer from the book the manner in which the author used it for his teaching (ibid.).

[13] § 27. If *a* and *b* are *Anzahlen*, and *A', A'', A'''*,... is a well-ordered set of the *Anzahl b*, while each of the elements *A* itself is a well-ordered set of the *Anzahl a*, then by dissolution of each *A* into its elements arises, in turn, a well-ordered set, the *Anzahl* of which is called the product of *a* and *b* and denoted by *ab* [...].

One must be aware that Felix Klein had emphatically polemicised against Meyer's approach in his *Elementarmathematik*. Set theory was a case for Klein where this theoretical development was too fresh, not yet accomplished, and even further from having matured to the point of having induced an intra-disciplinary process of integration and restructuration. Thus it was not yet elementarised: the concepts of set theory had not (yet) provide new elements for mathematics, hence Klein's polemic against Friedrich Meyer's schoolbook of 1885 (see Klein 2016, p. 289, note 181). Klein sharply criticised this schoolbook in his first edition of 1908 but softened his critique in subsequent editions (see ibid.).

6 Non-Euclidean Geometry in German Gymnasien

The last case concerns non-Euclidean geometry: in 1874, shortly after the first establishments of mathematical practices with the new geometries, which still met with strong resistance from many mathematicians and philosophers, a mathematics teacher at a Hamburg Gymnasium published a geometry textbook according to Bolyai's notion of absolute geometry.

In historiography, the decisive cause for accepting non-Euclidean geometry by a larger part of the mathematical community is attributed to the 1868 work of the Italian mathematician Eugenio Beltrami (1835–1900). In this work, the author presented Lobachevski's ideas in a geometric construction, establishing a description of the points inside a disc (Gray 1994, p. 881). The textbook presented here, hitherto unknown and not considered in historiography, reveals, however, another access to the new developments of geometry. The author, Hermann Wagner, having obtained a doctorate in mathematics and being a teacher of mathematics at a secondary school in Hamburg refers in fact not to Beltrami, but mainly to Bolyai and Riemann, mentioning Lobachevski briefly. The work he cites as a central reference for his approach is Riemann's famous masterpiece, published posthumously in 1867: *Über die Hypothesen, welche der Geometrie zu Grunde liegen.*

Wagner represents, similarly to Friedrich Meyer, the Prussian neo-humanist teacher, a profile socially recognised as that of a scholar, and with a professional performance that assumes to structure teaching in harmony with the coherence and rigour of its science. His book addressed two distinct audiences: the preface was written for his "Fachgenossen", the colleagues of his discipline, and the text itself was for school students of both the classic and modern streams. As at that time the teaching of geometry itself began in the "Quarta", corresponding to the third grade of the secondary school, and as this textbook was intended for the normal teaching of geometry, one can assume that it was written for students with an age of 12 years and over.

The entire book, and in particular its preface, shows that the author has no problem accepting the existence of different geometries; in fact, the intent is to make this recent breakthrough in mathematics accessible to students. Wagner explains at the beginning of the book that, on the one hand, geometry is one of the few sciences that has achieved a high degree of perfection, while, on the other, it is precisely its first foundations that still lack clarity and sufficient certainty. He highlights the absence of a definition of the straight line and the lack of a proof, which has been sought for centuries, of the axiom of parallels (equivalent to the theorem of the sum of angles in the triangle). The author notes with satisfaction that it has at last been proved in this century that all attempts at demonstrations of such an axiom should fail, attributing to Riemann the main merit of this result. For Wagner, this new approach in geometry represented a great epistemological significance: contrary to Kant's conceptions of geometry as an abstract science, of "reiner Anschauung", Riemann would have shown perfectly that the foundations of planimetry are grounded in experience and that, therefore, geometry presents itself as an "Erfahrungswissenschaft", an empirical science. This epistemological concept was of great importance for the author because it legitimates the introduction of the first concepts of geometry empirically—contrary to the dominant practice of his time (Wagner 1874, p. iii).

In fact, the existence of another quality of our "Raumform", the form of space, constitutes for Wagner a characteristic of the empirical. However, Wagner asserted that Bolyai had demonstrated the possibility of such a geometry in a "widerspruchsfrei" way, without contradictions, referring then to an axiomatic method.

The specific approach of his book stems from a book published in 1872 by Johannes Frischauf (1837–1924), professor of mathematics at the Austrian university of Graz: *Absolute Geometrie gemäß Johann Bol*yai. Although Frischauf incorrectly attributed the results of Wolfgang Bolyai to his father, who published his son's work as an appendix in a book, Wagner attributes to Frischauf's book the merit of having made "the genuine being of geometry" accessible to the general public. The author then proposed, as his task, to make the new concepts accessible to beginners, the students in secondary school. Since the traditional teaching of geometry was to present a logical and strictly related and deductible system of knowledge, the lack of clarity in the first foundations always presented a dilemma for teachers of mathematics (Wagner 1874, p. iv). As the task of his textbook, Wagner set out to begin with the simplest preconditions in planimetry, to demonstrate in the book only that which can be truly demonstrated, explaining that what cannot be demonstrated must be legitimised by experience.

In the rest of the preface, the author explained to his colleagues how he understands and teaches the basic concepts, point, line, plane and space, making one of the traditional choices for such an introduction: starting from the point as an infinitely small element of space and the line as movement of a point, etc. Of importance was his conceptualisation of the notion of direction, for the investigation of parallels in particular, leading to the concept of curvature in the case of spheres and the degree of curvature. Wagner hastened to assert to his colleagues that he did not intend to teach the concept of curvature to students of that age, but that he only explained it in the preface so as not to be accused of lack of understanding. Wagner also mentioned that the notion of angle presents problems still unresolved in its definition.

6.1 About the Contents of Wagner's Schoolbook

The structure of the book is very interesting—I have yet to encounter an analogue. The book has two parts:

- Absolute geometry, with three sections
- Euclidean geometry, with 5 sections

The absolute geometry sections deal with the straight line, the triangle composed of straight lines, and the sum of the angles in the triangle and the parallel lines. The "Euclidean" sections deal with the quadrilateral, the circle, the polygons (inscribed and circumscribed), proportionality, and the calculation of the content of the plane figures. Somehow, this structure corresponds even to the Elements of Euclid, since Euclid does not use the axiom of parallels at the beginning of the first book; the axiom happens to be used only from proposition 29 of book I onward.

Already in the introductory part of the book one remarks on an application of the empirical approach: given the key role of the concepts of congruence and equality, Wagner introduces them as "Erfahrungssatz"—a proposition legitimised by experience:

> All spatial quantities are independent of the place where they are (ibid., p.2).

Here Wagner also develops a definition of the straight line that proves to be sufficient for the definition of direction, angle, and finally of parallels in absolute geometry and in Euclidean geometry (ibid., pp. 3-4 and passim).

The third section is clearly the one of greatest interest here. In it, Wagner exposes and demonstrates all the results obtained since the eighteenth century on the sum of the angles in a triangle, including in particular the results of Legendre. He also adopts the use of Legendre's infinitely small, by "flattening" a triangle—namely by degenerating a triangle into a straight line (see Fig. 5.2).

More important are the theorems that state that the sum of the angles of a triangle cannot exceed two right angles; that from the sum of two right angles in *one* triangle, it follows that the sum would be the same in *each* triangle; and that the sum of each triangle is equal to two right angles or to a value smaller than this (see Fig. 5.3).

Following this is a well-argued discussion on the characteristics of parallel lines and the different cases of existence of only one parallel to a point in a line vertical

Fig. 5.2 Transforming a triangle into another with the same sum of angles, but in which the sum of two angles becomes "as small as one likes"

* § 73.

Lehrsatz: Jedes Dreieck läßt sich allmälich in ein anderes von gleicher Winkelsumme verwandeln, in welchem die Summe zweier Winkel beliebig klein ist.

Beweis: Um dies einzusehen, verwandle man (Fig. 35) $\triangle ACB$ zunächst in $\triangle ADB$, in welchem nach § 72:

$$\angle CAB = \angle DAB + \angle ADB.$$

Das so erhaltene Dreieck ADB behandle man wieder wie vorhin das Dreieck ACB. Man erhält alsdann ein Dreieck D_2AB (Fig. 36), in welchem $\angle D_2AB + \angle AD_2B$ nur noch so groß ist, wie $\angle DAB$.

Fig. 36.

Mit dem Dreiecke D_2AB kann man wieder ebenso verfahren und erhält, wenn man dies Verfahren beliebig oft wiederholt, eine Reihe von Dreiecken, in denen die Summe zweier Winkel beständig kleiner und kleiner wird. Nichts hindert, dies Verfahren bis ins Unbegrenzte fortgesetzt zu denken. So gelangt man schließlich zu Dreiecken, in welchen die Summe zweier Winkel kleiner ist, als jede noch so kleine Größe, so daß der dritte Winkel mit beliebiger Annäherung die ganze Winkelsumme des Dreiecks darstellt.

Fig. 5.3 Resuming the propositions about the sum of angles in a triangle

* § 75.

Zusatz: Die Summe der drei Winkel eines Dreiecks ist entweder gleich zwei Rechten oder kleiner als zwei Rechte.

to a straight line or of several parallels. The definitions, explanations and theorems are well explained and discussed, considering the level of the students.

7 Conclusion

Sufficient evidence for a productive, forward-looking function of mathematics teaching has been presented here. On the other hand, it is not possible to close our eyes to the fact that dogmatic, formalising impulses for the development of science have also emerged from the school in a fundamentalist exaggeration of the search for firm foundations.

Likewise, even institutionalised teaching of mathematics does not need to instigate production practices in mathematics. A striking example for such patterns deviating from the patterns presented in the second case is the medieval universities in Europe. While the quadrivium used to be part of teaching in the Arts Faculty, this teaching was not only relatively marginal, with regard to the trivium—the teachers of the quadrivium were freshly graduated *baccalaurei* who continued their studies in the higher faculties, without a specific qualification for these lecturers. Hence, institutionalisation, as a means for constituting a community developing their practices, has to be complemented by some professionalisation.

References

Sources

ETH-Bibliothek Zürich, 1858. Signatur Hs 797:2. Vorlesungsnachschrift von Heinrich Berchtold, Integral- und Differentialrechnung, Vorlesung von Julius Wilh. Richard Dedekind und von Heinrich Durège. Analytische Geometrie des Raumes, Vorlesung von Heinrich Durège.

Publications

Arbogast, Louis-François-Antoine. 1792. *Rapport et Projet de Décret sur la Composition des Livres Élémentaires Destines à l'Instruction Publique, Présentés à la Convention Nationale*. Paris: au Nom du Comité d'Instruction Publique.

Arnauld, Antoine. 1667. *Nouveaux Élémens de Géométrie*. Paris: Ch. Saveaux.

Belhoste, Bruno. 1998. Pour une réévaluation du rôle de l'enseignement dans l'histoire des mathématiques. *Revue d'Histoire des Mathématiques* 4: 289–304.

Bourbaki, Nicolas. 1948. L'architecture de mathématique. In *Les Grands Courants de la Pensée Mathématique*, ed. F. le Lionnais, 35–47. Paris: Blanchard.

Brandt, Christina. 2017. Kulturwissenschaften und Wissenschaftsgeschichte. In *Handbuch Wissenschaftsgeschichte*, ed. Marianne Sommer, Staffan Müller-Wille, and Carsten Reinhardt, 92–105. Stuttgart: Metzler.

Chevallard, Yves. 1985. *La Transposition Didactique—du Savoir Savant au Savoir Enseigné*. Grénoble: Éditons La Pensée Sauvage.

d'Alembert, Jean le Rond. 1755. Élémens des sciences. In *Encyclopédie ou Dictionnaire Raisonné des Sciences des Arts et des Métiers, Tome V*, 491–497. Paris: Briasson.

Dauben, Joseph, and Christoph Scriba. 2002. *Writing the History of Mathematics*. Basel, Boston, Berlin: Birkhäuser.

Dedekind, Richard. 1872. *Stetigkeit und irrationale Zahlen*. Braunschweig: Vieweg.

Destutt de Tracy, Antoine L.-C. 1801. *Projet d'Eléments d'Idéologie [Project of a treatise of ideology]*. Paris: Didot.

Fávero, Maria de Lourdes A., and Sonia de Castro Lopes. 2009. *A Universidade do Distrito Federal (1935-1939). Um Projeto Além de seu Tempo*. Brasilia: Liber Livro Editora.

Frischauf, Johannes, ed. 1872. *Absolute Geometrie Nach Johann Bolyai*. Leipzig: Teubner.

Gray, Jeremy. 1994. Euclidean and non-Euclidean geometry. In *Companion Encyclopedia of the History and Philosophy of the Mathematical Sciences*, Volume 2, ed. Ivor Grattan-Guinness, 877–886. London and New York: Routledge.

Hoffmann, J.C.V. 1899. Zum Andenken an Dr. Friedrich Meyer. *Zeitschrift für Mathematischen und Naturwissenschaftlichen Unterricht* 30: 551–553.

Klein, Felix. 2016. *Elementary Mathematics from a Higher Standpoint. Arithmetic, Algebra, Analysis*. vol. 1. Trans. G. Schubring. Springer: Berlin & Heidelberg.

Lietzmann, Walther. 1909. *Stoff und Methode im Mathematischen Unterricht*. Leipzig & Berlin: Teubner.

Lorey, Wilhelm. 1915. Der 70. Geburtstag des Mathematikers Georg Cantor. *Zeitschrift für Mathematischen und Naturwissenschaftlichen Unterricht* 46: 268–274.

Metternich, Mathias. 1818. *Die Reine und Angewandte Zahlenlehre, für Lehrer und Studirende*. Coblenz: Neue Gelehrte Buchhandlung.

Meyer, Friedrich. 1885. *Elemente der Arithmetik und Algebra. Zweite Auflage*. Halle: H. W. Schmidt.

———. 1891. *Mitteilungen aus Dem Mathematischen Lehrplan dos Gymnasiums*. Halle: Schulprogramm des Stadtgymnasiums zu Halle a. S.

Pereira, Vinicius Mendes Couto. 2017. *O Desenvolvimento da Análise no Brasil—um Caminho Sobre o Surgimento de Uma Comunidade Matemática*. PhD thesis, Universidade Federal do Rio de Janeiro.

Purkert, Walter. 1986. *Georg Cantor und die Antinomien der Mengenlehre*. Leipzig: Preprint Naturwissenschaftliches-Theoretisches Zentrum Kart·Marx-Universität.

Purkert, Walter, and Hans Joachim Ilgauds. 1987. *Georg Cantor 1845-1918*. Basel, Boston & Stuttgart: Birkhäuser.

Ramus, Petrus. 1569. *Scholarum Mathematicarum Libri Unus et Triginta*. Basileae: Episcopius.

Schubring, Gert. 1982. Ansätze zur Begründung theoretischer Terme in der Mathematik - Die Theorie des Unendlichen bei Johann Schultz (1739–1805), *Historia Mathematica*, *9*, 441–484.

Schubring, Gert. 1987. On the methodology of analysing historical textbooks: Lacroix as textbook author. *For the Learning of Mathematics* 7(3): 41–51. ("Errata", ibid., 1988, *8: 2*, 51).

———. 1988. Introduction: "Un savant des lumières—Un livre élémentaire pour la république". In *CONDORCET—Moyens d'Apprendre à Compter Sûrement et Avec Facilité. Appareil Critique—Études, Notes, Commentaires, Bibliographie*, 157–175. Paris: ACL-éditions.

———. 1990. Zur strukturellen Entwicklung der Mathematik an den deutschen Hochschulen 1800-1945. In *Mathematische Institute in Deutschland 1800-1945*, ed. Winfried Scharlau, 264–278. Braunschweig: Vieweg.

———. 1991a. *Die Entstehung des Mathematiklehrerberufs im 19. Jahrhundert. Studien und Materialien zum Prozeß der Professionalisierung in Preußen (1810-1870)*. Zweite, korrigierte und ergänzte Auflage. Weinheim: Deutscher Studien Verlag.

———. 1991b. Spezialschulmodell versus Universitätsmodell: Die Institutionalisierung von Forschung. In *'Einsamkeit und Freiheit' neu Besichtigt*, ed. G. Schubring, 276–326. Stuttgart: Franz Steiner Verlag.

———. 2001. Production Mathématique, Enseignement et Communication. Remarques sur la note de Bruno Belhoste, "Pour une réévaluation du rôle de l'enseignement dans l'histoire des mathématiques" parue dans la RHM 4(1998), p. 289-304. *Revue d'Histoire des Mathématiques* 7: 295–305.

———. 2011. Conceptions for relating the evolution of mathematical concepts to mathematics learning—epistemology, history, and semiotics interacting. *Educational Studies in Mathematics* 77 (1): 79–104.

Seebold, Carl. 1821. *Elemente der Arithmetik*. Giessen.

Silva, Circe Mary Silva da. 1991. Die Gründung der mathematischen Fakultät 1772 in Coimbra - ein Beginn fachlicher Spezialisierung. In: *'Einsamkeit und Freiheit' neu besichtigt,* Gert Schubring, ed., 50–56. Stuttgart: Franz Steiner.

Sommer, Marianne, Staffan Müller-Wille, and Carsten Reinhardt, eds. 2017. *Handbuch Wissenschaftsgeschichte*. Stuttgart: Metzler.

Tait, William W. 2000. Cantor's *Grundlagen* and the paradoxes of set theory. In *Between Logic and Intuition: Essays in Honor of Charles Parsons*, ed. G. Sher and R. Tieszen, 269–290. Cambridge: Cambridge University Press.

Trouvé, Alain. 2008. *La Notion de Savoir Élémentaire à l'école. Doctrines et Enjeux*. Paris: L'Harmattan.

Wagner, Hermann. 1874. *Lehrbuch der Ebenen Geometrie. Nach Grundsätzen Bolyais für Gymnasien und Realschulen bearbeitet*. Hamburg: Lucas Gräfe.

Chapter 6
A Multiple Perspective Approach to History of Mathematics: Mathematical Programming and Rashevsky's Early Development of Mathematical Biology in the Twentieth Century

Tinne Hoff Kjeldsen

Abstract A multiple perspective approach to history of mathematics is introduced. It is discussed how such an approach can be used to explore and shed light on relationships between developments in mathematics and the conditions of those developments. The approach will be illustrated by two episodes from the history of applied mathematics from the twentieth century: the development of mathematical programming and the significance of World War II, and Nicolas Rashevsky's early attempts to develop mathematical biology in the 1930s in the USA and issues of engagement with interdisciplinary research. The importance that the historical actors attached to education and teaching for the development of these new fields is discussed, drawing attention to education and teaching as relevant and significant perspectives for understanding historical developments in mathematics.

Keywords Education · Historiography · History of twentieth century mathematics · Interdisciplinary research · Linear and nonlinear programming · Mathematical biology · Military-university complex · Multiple perspective approach to history of mathematics · Nicolas Rashevsky · Operations research

1 Introduction

In this chapter, I introduce and discuss a multiple perspective approach to history of mathematics,[1] and how it can be used to explore and shed light on relationships between developments of mathematics and conditions of its developments. I present

[1] I have borrowed and translated the term "a multiple perspective approach" from the historian Bernard Eric Jensen's term "en flerperspektivist tilgang" from his book "historie—livsverden og

T. H. Kjeldsen (✉)
Department of Mathematical Sciences, University of Copenhagen, Copenhagen, Denmark
e-mail: thk@math.ku.dk

G. Schubring (ed.), *Interfaces between Mathematical Practices and Mathematical Education*, International Studies in the History of Mathematics and its Teaching,
https://doi.org/10.1007/978-3-030-01617-3_6

two episodes from the history of applied mathematics from the twentieth century to illustrate this historiographical approach: (1) the development of mathematical programming and the significance of World War II and (2) Nicolas Rashevsky's early attempts to develop mathematical biology in the 1930s in the USA, related to issues of engagement with interdisciplinary research. The historical actors associated with these two episodes attached special importance to education and teaching for the development of their fields. Some of their concerns are discussed to draw attention to education and teaching as relevant and significant perspectives for understanding historical developments in mathematics.

2 A Multiple Perspectives Approach to History of Mathematics

As it is stated in the introduction, this book consists partly of chapters that were presented in the workshop organized by Gert Schubring at the Fourth International Meeting of the Association for the Philosophy of Mathematical Practices (APMP).[2] The initial conception of the workshop was inspired by Schubring's (2001) discussion of a note by Bruno Belhoste (1998) that in a broader sense addressed the question of how mathematics is produced and the impact of particular conditions for its production, emphasizing that historians should not, a priori, separate the production of mathematics from the conditions of its production.

Belhoste (1998, p. 289) was criticizing what he considered to be the traditional view at that time that the development of mathematics was entirely autonomous. Belhoste was writing two decades ago, at a time when the so-called internalism/externalism debate was still going on in the historiography of mathematics. Joan Richard (1995) for example, identified the debate between internalism and externalism as *the* critical problem for history of mathematics at the turn of the twenty-first century. At the same time, Moritz Epple (2000) discussed this contentious issue in his paper "Genies, Ideen, Institutionen, mathematische Werkstätten: Formen der Mathematikgeschichte." He described the controversy as a disagreement among historians of mathematics about the decisive (he called them causal) factors that affect the development of mathematics. In a purely internalist approach, the decisive factors responsible for the development of mathematics are sought only within mathematics, which is itself considered to be immune from influences from outside. By now, however, it is generally recognized by historians of mathematics that factors external to mathematics play significant roles in the historical development of mathematics as well.

fag" (History—lifeworld and discipline), Jensen (2003). I have also discussed this in Kjeldsen (2009, 2012).

[2] The conference took place in Salvador, Brazil in October 2017, http://www.philmathpractice.org/2016/06/11/fourth-meeting-of-the-apmp-october-23-27-2017-salvador-da-bahia-brazil/

In the following, I will introduce a multiple perspective approach to history of mathematics based on an action-oriented conception of history of mathematics. It is an approach that moves beyond this distinction between internalism and externalism and opens for exploring interactions between mathematics and its broader social context. In such a conception, mathematics develops through people's activities in their life and work, conditioned and modified by the past, the present, and expectations for the future. People's development, use, displays, teaching, etc. of mathematics arise in particular times and places in various concrete contexts, scientific as well as non-scientific, and their dissemination of mathematical ideas, methods, and knowledge depend on among others a common language and a shared culture.

By taking people's motivation and their actions as points of departure, paying attention to intended and unintended consequences of their actions, historical investigations get rooted in people's concrete projects through which connections and relations can be explored and unfolded, projects that can be studied from various points of observation or perspectives. Specific mathematical activities and/or circumstances for these at certain times and places can be investigated and analyzed through various lenses, for example, from practices of mathematics, from applications of mathematics, from positions about the "nature of mathematics," from interdisciplinary perspectives, from sociological angles such as institutions and/or personal networks, from philosophies and practices of teaching and education, from religious beliefs, from gender parity, and so forth. Such an approach will not likely identify one decisive factor responsible for the development of mathematics, but instead it will point towards different factors acting together in a web of events that had an influence, for example, on the production of a particular piece of mathematical knowledge, on changes in mathematical practices, on the move of mathematics into other areas, or the development of views on mathematics. It is an open approach to history, where the interesting perspectives are determined by the research questions.

This kind of openness leaves room for various presentations of historical episodes. Historical processes may be perceived differently from each of the multiplicity of viewpoints. This raises the questions of reliability and validity. Reliability of the historian's analyses and conclusions depends on the relation between the research questions, the perspectives of the actors that are taken into account, their motivations and actions, and the material and sources used by the historian, and finally the tools, theories, and methodological frameworks that the historian uses in the analyses (see Fig. 6.1).

Research questions are governed by the curiosity of historians, issues historians wonder about, and processes of change that they are seeking to explain, which again are determined by their perspectives. Hence, there is another layer on top (see Fig. 6.2) that is connected to the question of validity in the sense of relevance of the historical investigation. The answers depend on relationships between historians' views, the matters of concern under investigation and the intended audiences (the readers). One way to begin such a discussion is for historians to declare their positions in relation to the subjects/episodes/histories they are writing about and the

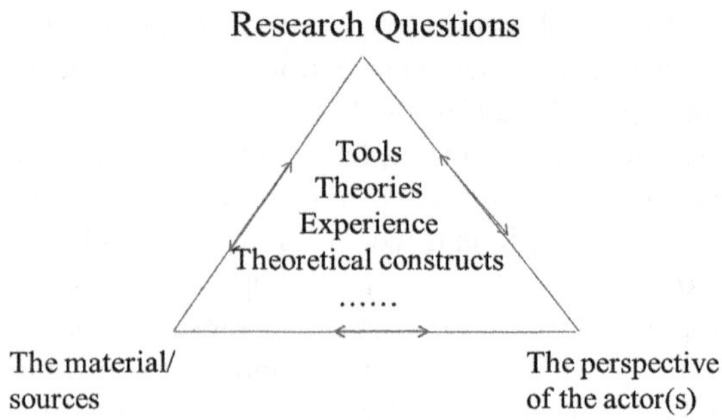

Fig. 6.1 Relation between research questions, the perspective of actors and sources

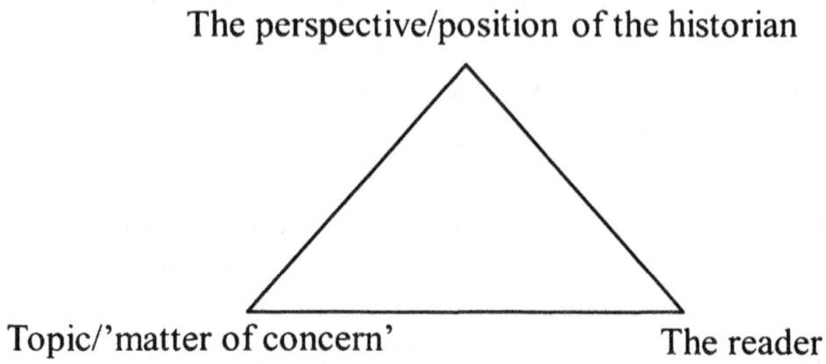

Fig. 6.2 Relation between the historian, the topic, and the reader

approaches they are using, so, as phrased by the historian Søren Mørch (2010, p. 503) "the reader can relate to the history writing he or she is being dragged into."

In this conception of history, the actors' perspectives are crucial for understanding historical processes of development and change. The relevant perspective(s) depend on the historian's research questions, which again depend on the historian's perspective.

3 Two Episodes: Mathematical Programming and Mathematical Biology

In the following, I will discuss two episodes from twentieth-century mathematics. The focus will be on emphasizing the historiographical approach discussed above in exploring interactions between development of mathematics and conditions of its development. The first episode concerns the development of mathematical programming in the USA and the significance of World War II. The second episode involves Nicolas Rashevsky's attempt in the 1930s in the USA to apply mathematics to biology as a method of explaining biological processes and issues of engagement with interdisciplinary research. The presentations below draw on my research in history of mathematics related to these two episodes.

3.1 Mathematical Programming in the USA and the Significance of World War II

In the Mathematics Subject Classification system[3], Operations Research and Mathematical Programming are placed together as subject no. 90, with linear and nonlinear programming as sub-areas.[4] In the USA, these fields emerged and developed into research areas of mathematics in connection with the Second World War.[5] Linear programming grew out of a logistic problem in the US Air force that George B. Dantzig (1914–2005), among others, was working on during and after the war, whereas nonlinear programming began as an attempt by Albert W. Tucker (1905–1995) and Harold W. Kuhn (1925–2014) at the Mathematics Department at Princeton University to generalize the duality theorem for linear programming to the nonlinear case.

My historical investigations of this episode in the history of mathematics were stimulated by a curiosity about how ideas of duality emerged in linear programming. What role did they play for the development of nonlinear programming on the one hand, and what role did the military play for the emergence and development of mathematical programming on the other? What influence did these factors have on the establishment of mathematical programming as a research area in mathematics in academia? The source material included mathematical papers, letters between actors, their reminiscences, and interviews with actors, as well as secondary literature on the military-university complex in the USA during and after WWII. My historical investigation was guided by the underlying issue of driving forces in the history of mathematics, from the perspective of the actors' motivation to engage in and produce this kind of mathematics as well as the conditions for this engagement and development of mathematics. The analyses show that driving forces that are both internal and external to mathematics were present and interacted in crucial ways to influence mathematicians' development of the field.

The duality theorem is a key result in the mathematical theory of linear programming. It states that if a linear programming problem has a finite optimal solution so does the dual one and the optimal values are the same. Dantzig (1963), in his textbook on linear programming, ascribes the duality theorem to John von Neumann. Apparently, this came as a surprise to Tucker because he, Kuhn, and David Gale (1921–2008) had published the first existence and duality theorems for linear programming in 1949. Dantzig (1982, p. 45) elaborates his attribution in his recollections, where he mentions a conversation he had had with Tucker, where Tucker had asked him why he ascribed duality to von Neumann and not to Tucker and his

[3] https://mathscinet.ams.org/msc/pdfs/classifications2010.pdf

[4] Mathematical programming is dealing with finite dimensional optimization under inequality constraints.

[5] The Russian mathematician and economist Leonid V. Kantorovich published the paper *Mathematical Methods of Organizing and Planning Production* containing similar ideas in 1939. For an English translation, see Kantorovich (1960). See also Leifman (1990).

group, to which Dantzig should have answered: "Because he [von Neumann] was the first to show it to me" (Dantzig 1982, p. 45). According to Dantzig, von Neumann did so in a meeting between the two of them in the fall of 1947, and in his recollections, he goes on to explain how he remembered what went on at that meeting:

> I remember trying to describe to von Neumann, as I would to an ordinary mortal, the Air Force problem. I began with the formulation of the linear programming model in terms of activities and items, etc. Von Neumann did something which I believe was uncharacteristic of him. 'Get to the point', he said impatiently. [...] I said to myself 'O.K., if he wants a quicky, then that's what he'll get'. In under one minute I slapped the geometry and the algebraic version of the problem on the blackboard. Von Neumann stood up and said 'Oh that!'. That for the next hour and a half, he proceeded to give me a lecture on the mathematical theory of linear programs. At one point seeing me sitting there with my eyes popping and my mouth open (after all I had searched the literature and found nothing), von Neumann said: 'I don't want you to think I am pulling all this out of my sleeve on the spur of the moment like a magician. I have just recently completed a book with Oscar Morgenstern on the theory of games. What I am doing is conjecturing that the two problems are equivalent. The theory that I am outlining for your problem is an analogue to the one we have developed for games'. Thus I learned about Farkas's Lemma, and about duality for the first time. (Dantzig 1982, p. 45)

The account of this meeting between the two is Dantzig's reconstruction of the event—how he remembered it when he wrote it down. However, it is interesting because it displays a collection of various kinds of situated links in a concrete social and communicative situation at Princeton. On a personal level, there is a link between two mathematicians. On an institutional level, there is a link between the military where Dantzig was employed to work on what he referred to as "the Air Force problem" in the quote above, and academia where von Neumann was employed as a professor at the Institute for Advanced Study at Princeton University. On a mathematical level, a link between the algebraic version of the Air Force problem (to minimize a linear form subject to linear inequalities) and the mathematical theory of games was realized. A transmission of knowledge across various kinds of contextual boundaries took place, conditioned by the mathematical interests of the two participants, their work situation, and their roles in the meeting as well as the circumstances of the post-war period in the USA.

Dantzig was hired by the Air Forces in 1941 as part of the scientific mobilization to work on what they called "programming planning methods" which were schedules or programs, as they were called, for huge logistic planning. During the war, Dantzig (1982, 1991) trained Air Force staff on how to calculate such programs. By then he had a master's degree in mathematics from University of Maryland, and 2 years of experience as a junior statistician at the U.S. Bureau of Labor Statistics.

After the war, he finished his PhD study and from 1946 he was re-employed by the military, where he continued to work on the programming problem and methods for speeding up the computation of programs. During the war, the objective had been to compute consistent programs but soon after the war the computer was invented and with it the possibility of calculating optimum programs. Together with the economist Marshall Wood, who was an expert on military programming procedures, Dantzig began to work on what they called scientific computation of optimum programs. In a paper published in the journal *Econometrica* in 1949, the two of them discussed the programming of interdependent activities:

> We seek to determine that program which will, in some sense, most nearly accomplish objectives without exceeding stated resources limitations. So far as is known, there is so far no satisfactory procedure for solution of the type of problem. (Dantzig and Wood 1949, p. 195)

Formulated in mathematical terms, it gave rise to the problem of minimizing a linear form subject to linear equations and inequalities, which is a linear programming problem as it is most often formulated in mathematical textbooks today.

According to Dantzig, von Neumann recognized a connection between Dantzig's algebraic version of the programming problem and the theory of games at their meeting, and introduced Dantzig to Farkas's lemma and duality. How could he do that? After all, Dantzig claimed that he had searched the literature and found nothing. Looking into von Neumann's work in mathematics, we find that he had derived his first results in game theory 20 years earlier. In 1928, he published the paper "Zur Theorie der Gesellschaftsspiele" in which he asked the question, given *n* players who are playing a game of strategy, "how must one of [them] play in order to achieve a most advantageous result?"[6] He characterized this question as "the principal problem of classical economics: how is the absolutely selfish "homo economicus" going to act under given external circumstances?"[7] This interpretation laid the foundation for his and Morgenstern's work on economic theory 15 years later, when in Chap. 1 of their book *Theory of Games and Economic Behavior* they wrote that:

> It will then become apparent that there is not only nothing artificial in establishing this relationship [between theory of games and economic theory] but that on the contrary this theory of games of strategy is the proper instrument with which to develop a theory of economic behavior. (von Neumann and Morgenstern 1944, p. 1–2)

Hence, with respect to applications, von Neumann considered the mathematical theory of games of strategy as the way to mathematize economic behavior from the very beginning. In the 1928 paper, he proved the existence of optimal (mixed-)strategies, the so-called minimax theorem, for finite two-person zero-sum games, that is games with two participants where their winnings and loses add up to zero.

An analysis of von Neumann's proof of the minimax theorem and his encounter with the theorem leading up to his joint work with Morgenstern, which he referred to in Dantzig's recollection of their meeting, can offer some insights into how he could make the connection between the Air Force problem and game theory.[8] His first proof from 1928 has been called a "a tour de force."[9] He first proved that a pair of optimal strategies, one for each player, constitutes a saddle point for the payoff

[6] von Neumann (1928, p. 295).

[7] von Neumann (1928, p. 295, note 2).

[8] For an analysis of von Neumann's work and encounter with the minimax theorem in different contexts, see Kjeldsen (2001).

[9] Heims (1980, p. 91).

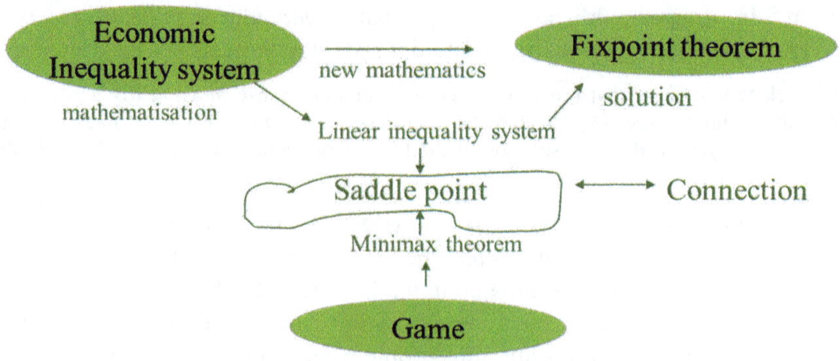

Fig. 6.3 Analysis of connections in von Neumann's 1928 paper on game theory and his-1937 paper on the mathematization of an economic problem

function associated with the game, and he used this result to prove the minimax theorem by proving the existence of such a saddle point.

Four years later, in 1932, von Neumann gave a talk at the mathematics seminar at Princeton. The talk was published 5 years later at the request of Karl Menger in a paper with the title "Über ein ökonomisches Gleichungssystem und eine Verallgemeinerung des Brouwerschen Fixpunktsatzes."[10] In this chapter, the minimax theorem reappeared, however in another guise. Von Neumann considered a general economic situation where n goods are produced by m processes, and asked the question: "Which processes will be used (as "profitable") and what prices of the goods will obtain?" His mathematization resulted in a system of six linear inequalities expressing relations between the intensities of the processes and the prices of the goods. To prove the existence of a solution to the inequality system, he transformed the problem into a problem of a saddle point for a certain function, and proved a lemma that he called a generalization of Brouwer's fixed-point theorem. He then used this lemma to prove the existence of a saddle point. As in his 1928-paper, he turned the problem into the problem of proving the existence of a saddle point for a certain function. That von Neumann was aware of this connection between existence of solutions to linear inequality systems and optimal strategies for two-person zero-sum games in 1937, is evident from the following quote:

> the question whether our problem (system of linear inequalities) has a solution is oddly connected with that of a problem occurring in the Theory of Games dealt with elsewhere. (von Neumann 1937, p. 79, footnote 2)

In these works of von Neumann, the application of mathematics and the derivation of new mathematical results interacted in different ways and became connected through the underlying saddle value problem (see Fig. 6.3).

Seven years later in 1944, von Neumann and Morgenstern published *Theory of Games and Economic Behavior*. Here they derived the minimax theorem as a simple

[10] von Neumann (1937). An English translation was published in 1945 under the title "Model of General economic equilibrium."

consequence of a theorem about systems of linear inequalities, the so-called Theorem of the Alternative for Matrices. They based their proof on a theorem of separating hyperplanes, which is an essential tool in convex analysis. In this book, they proved the minimax theorem within the mathematical theories of systems of linear inequalities and convexity inspired by a French mathematician, Jean Ville, who, in this development of the proof of the minimax theorem, took one of the first steps towards a mathematical context of convex analysis. According to Morgenstern, it was Ville's proof that became the direct inspiration for von Neumann and Morgenstern's algebraic version of the proof of the minimax theorem in the 1944-book.[11]

The minimax theorem in game theory had a short aftermath. A decade later, in 1953, the French mathematician Maurice Fréchet had three notes of Emile Borel from the period 1921 to 1927 translated into English. In connection with his work on probability, Borel had discussed gambling, and in this connection also games of strategy. In the introduction to the translation of Borel's notes, Fréchet argued for the importance of Borel's work:

> It was only relatively recently that I began to occupy myself with the theory of probability and its applications, which explains why the notes that Émile Borel … published between 1921 and 1927 on the theory of psychological games escaped my attention. It was chance to begin with … because, in the extensive literature devoted to this theory [game theory] and its applications in recent years, references to earlier work do not lead back, in general, further than to the important paper published in 1928 by Professor von Neumann. But, in reading these notes of Borel's I discovered that in this domain, as in so many others, Borel had been an initiator. (Fréchet 1953, p. 95)

This introduction by Fréchet caused a brief priority debate between von Neumann and Fréchet in which von Neumann pointed out that Borel failed to prove the minimax theorem and in von Neumann's opinion "there could be no theory of games on these bases without that theorem … I felt that there was nothing worth publishing until the 'minimax theorem' was proved"[12] to which Fréchet claimed that:

> Again, it may be mentioned, that even if Borel had, before von Neumann, established the minimax theorem in its full generality; the profound originality of Borel's notes would not have been augmented nor even touched from the economic point of view. He would not thereby have even enriched the set of properly mathematical discoveries for which Borel has acquired a world-wide reputation. He would have, like von Neumann, simply entered an open door. […] the same theorem and even more general theorems had been independently demonstrated by several authors well before the notes of Borel and the first paper of von Neumann. (Fréchet 1953, p. 122)

The proofs Fréchet is referring to are proofs of theorems similar to von Neumann and Morgenstern's "Alternatives for Matrices," i.e., theorems about solutions to systems of linear inequalities by Minkowski, Farkas, Stiemke, and Weyl. Dantzig mentioned that he learned about Farkas's lemma on systems on linear inequalities in his

---

[11] See Kuhn and Tucker (1958, p. 116) and Rellstab (1992, p. 87). For an analysis of Ville's proof, see Kjeldsen (2001).

[12] von Neumann and Fréchet (1953, pp. 124–125).

first meeting with von Neumann. The Hungarian Gyula (Julius) Farkas (1847–1930) presented this lemma in a longer paper on "A theory of linear inequalities," published in 1901. Leaving Fréchet's motivations for raising the priority question aside, the discussion can be used to address an internal aspect of the issue of the production of mathematics and its conditions.

Farkas developed his lemma in an effort to establish an algebraic foundation for the extension of Lagrange's multiplier rule to Fourier's inequality principle in mechanics. The 1901 paper in which he presented a theory of linear inequality systems is essentially a reworked version of ideas presented in earlier papers on an algebraic foundation for the principle (Farkas 1895, 1897, 1899). In Fréchet's evaluation, half a decade later, the minimax theorem and Farkas's lemma are considered to be the same theorem in the sense that the minimax theorem did not "enrich the set of properly mathematical discoveries." Instead, if we compare and evaluate the work of Farkas and von Neumann from the process of their research a different picture emerges. We have identified two strands of research: (1) Farkas's that led him from his concern of extending Lagrange's multiplier rule to Fourier's inequality principle in mechanics to an algebraic theory of linear inequalities; and (2) von Neumann's that led him to the presentation of the theory of games and economic behavior within the theory of linear inequalities and convexity theory. As the analyses have shown, these two strands of research were very different, and they led to different outputs as illustrated in Fig. 6.4.

Fréchet's interpretation in 1953 of the minimax theorem as a simple consequence of Farkas's lemma was taken from a perspective of the universality of mathematics. However from a historical "developing of mathematics" point of view within the historiographical approach outlined in Sect. 2, Fréchet's evaluation of von Neumann's minimax theorem as already part of the body of mathematical knowledge due to Farkas's lemma, and thereby adding nothing new, collapses the historical process. There are no links to questions of games of strategy in Farkas's work, and von Neumann did not develop game theory on the basis of linear inequalities in

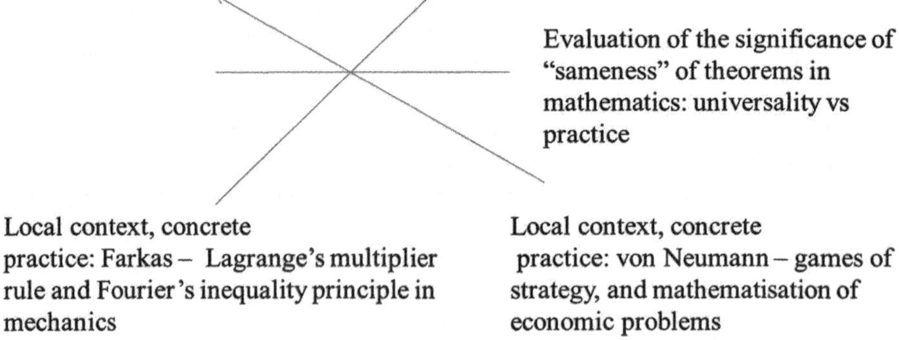

Fig. 6.4 Two strands of research situated in two mathematicians' different projects

the beginning. These insights only followed later, when he encountered the minimax theorem in various mathematical and applicational contexts, as outlined above, see also Kjeldsen (2001). The history of von Neumann's development and conception of the minimax theorem shows that it was far from trivial and took a larger effort to realize the connection between solutions of systems of linear inequalities and the existence of optimal strategies for two-person zero-sum games. The mathematics a mathematician or group of mathematicians can develop is conditioned by their practice, their mathematical "workshop," their research agenda and motivation, their capabilities, and the circumstances of their lives.

According to Dantzig, von Neumann made the connection between game theory and the Air Force problem at a meeting in Princeton, a meeting that was conditioned by the scientific mobilization in the USA during the Second World War. Von Neumann held a lot of consultant jobs for the military in the post-war period, and it was in this capacity that he met with Dantzig. Dantzig on his side was advised to consult with von Neumann regarding how to solve the Air Force problem. Their meeting was a consequence of the organization and mobilization of (civilian) scientists in the USA during the war, where mathematicians (scientists) working on military problems became connected with mathematicians (scientists) working in academia. Channels of communication between the two institutions where established in the post-war period when civilian scientists began serving as consultants for the military and funding for research in academic institutions was increased.

Von Neumann connected the Air Force programming problem with game theory, thereby placing the programming problem within a mathematical theory. It then crossed the boundary into academia through the request and financial support of the Office of Naval Research (ONR), initially for a summer project in 1948 at the mathematics department at Princeton with Tucker as the principal investigator. From Dantzig's recollections it seems as if Tucker became involved almost by coincidence. But the ONR supported project on logistics was not a coincidence; it was the result of a well-thought plan for promoting and controlling research in the post-war period. Mina Rees (1902–1997), who had served as technical assistant to Warren Weaver, the head of the Applied Mathematics Panel during the war, became the head of the mathematics section of ONR after the war. She described how she remembered it in 1977:

> [...] when, in the late 1940's the staff of our office became aware that some mathematical results obtained by George Dantzig, [...] could be used by the Navy to reduce the burdensome costs of their logistics operations, the possibilities were pointed out to the Deputy Chief of Naval Operations for Logistics. His enthusiasm for the possibilities presented by these results was so great that he called together all those senior officers who had anything to do with logistics, as well as their civilian counterparts, to hear what we always referred to as a "presentation". The outcome of this meeting was the establishment in the Office of Naval Research of a separate Logistics Branch with a separate research program. This has proved to be a most successful activity of the Mathematics Division of ONR, both in its usefulness to the Navy, and in its impact on industry and the universities. (Rees 1977, p. 111)

Albert W. Tucker worked together with Harold W. Kuhn and David Gale on the ONR project during the summer of 1948. The aim of the project was to explore further the connection between linear programming problems and game theory, and to do research in the underlying mathematical theories of linear inequalities and convexity.

The first results of their work are reported in the paper "Linear programming and the Theory of Games" which they presented the following summer at what is now thought of as the first conference on linear programming. The very tight connection to the military and the role of the military-university complex is reflected in the list of participants and their sponsors. The conference itself and most of the research were supported by the RAND Corporation or the U.S. Department of the Air Force, and most of the university contributions were done under contract with ONR.

In their chapter, Tucker, Gale, and Kuhn proved the duality theorem and existence theorems—not for the "basic" linear programming problem but for a generalized "matrix" problem that has the ordinary "basic" problem as a special case. They also gave a new proof of the minimax theorem and showed that the optimal strategies for a two-person zero-sum game constitute a solution to the corresponding "basic" linear programming problem and its dual. A dual problem to a linear programming problem is itself a linear programming problem on the same set of data. If the primal linear programming problem is a maximization problem, the dual is a minimization problem. The duality theorem states that if one of the problems has a finite optimal solution so does the other one, and the corresponding optimal values are the same. Tucker, Kuhn, and Gale's proceedings paper entail the first rigorous proof of the duality theorem in linear programming. Here, we see another effect of the military-university complex. Kuhn, Gale, and Tucker treated the linear programming problem as a mathematical research field, instead of just working on the "basic" linear programming problem they generalized it to a "matrix" version without any consideration of the applicability. This approach is typically for basic research in pure mathematics as it is conducted in academia.

The duality theorem caught the further interest of Tucker who intended to take it to the next level. He asked Gale and Kuhn whether they wanted to continue working on the project to see if they could extend the duality result to the nonlinear case (Kuhn 1976). Gale declined but Kuhn went along with Tucker and in 1950 they presented their paper "Nonlinear Programming." They did not succeed in proving a duality result for nonlinear programming, but they did prove what immediately became known as the Kuhn-Tucker conditions for the existence of a solution to a nonlinear programming problem. The significance of duality and the connection to game theory for Kuhn's and Tucker's work into nonlinear programming is evident from their treatment of the nonlinear case. They reformulated the nonlinear extremum problem under nonlinear inequality constraints as a saddle value problem for the corresponding Lagrangian, because, as they stated in the introduction of their "Nonlinear Programming" paper:

> This problem (a linear programming problem) can be transformed as follows into an equivalent saddle value (minimax) problem ... a saddle point ... provides a solution for a related zero sum two person game. The bilinear symmetry of [the saddle value problem] in [the two variables] yields the characteristic duality of linear programming. (Kuhn and Tucker 1950, p. 481)

The duality results for linear programming came out of the connection to von Neumann's work on two-person zero-sum games and it spurred the further work of Kuhn and Tucker into nonlinear programming[13]—through mathematical connections that were realized and through funding possibilities that were created, both of which were opportunities made possible by WWII and the following post-war organization of science support in the USA.

3.2 *Beginning of Mathematical Biology and Interdisciplinary Research*

The second example deals with Nicolas Rashevsky's early work in the beginning of mathematical biology in the USA in the 1930s. He advocated for a change in scientific practice in biology and he had a hard time getting through to the biologists of his time. The following presentation of Rashevsky's early work in mathematical biology has been guided by the underlying question of how scientists argued for and viewed the epistemic role of mathematics and the emergence of modeling in their efforts to mathematize other areas and how these efforts were perceived by practitioners from these areas, here exemplified by Nicolas Rashevsky's early work in mathematical biology. His scientific work and philosophical ideas have been analyzed with a focus on his motivation, perception of and goals for the role of mathematics as a scientific method in biology, and why his ideas clashed with the biologists' views from the perspective of interdisciplinary research.[14]

Nicolas Rashevsky (1899–1972) was born in Chernigov in Ukraine. He was trained as a theoretical physicist at the University of Kiev. He immigrated to the USA in 1924, to Pittsburgh where he worked as a research physicist at Westinghouse Electric and Manufacturing Company for 10 years. While at Westinghouse he published papers on the dynamics of colloid particles and division of droplets, and he made an analogy to the process of cell division:

> Is not the most important biological process, the division of cells, considered in purely physical terms, nothing but a spontaneous dispersion of a highly complicated built drop, or better, drop system? (Rashevsky 1928, 568)

Mathematics and statistics were being used in biology at that time. For instance, Karl Pearson founded the journal *Biometrika* in 1901, in the 1920s Vito Volterra in Italy and Alfred Lotka in the USA worked on population dynamics, and William O. Kermack and Anderson G. McKendrick in Scotland published their joint work "A Contribution to the Mathematical Theory of Epidemics" in 1927, to name but a few examples. However, according to Rashevsky, writing about mathematical biophysics in *Nature* in 1935, "no systematic attempt to create a mathematical biology has

---

[13] See Kjeldsen (2000).

[14] See Kjeldsen and Blomhøj (2013) and Kjeldsen (2017) where Rashevsky's early work on cell division is discussed in relation to its value for teaching students to reflect upon mathematical modeling and the modeling process.

been made" (Rashevsky 1935, p. 528). At that point in time, he was at Chicago University on a fellowship from the Rockefeller Foundation.[15]

Warren Weaver was by then head of the Rockefeller Foundation's division of the natural sciences. Mina Rees, in a biographical memoir of Weaver writes that when he was asked by the trustees at an interview in the fall of 1931 about his ideas for the Rockefeller program for the support of science he urged them to "undertake a long-range program of support of quantitative biology – a program that would seek to apply to outstanding problems of biology some of the methods and machines that had been so successful in the physical sciences." (Rees 1987, p. 500). In January 1932, Weaver was elected director for the Natural Sciences of the Foundation, and he proposed that the Foundation's science program should take an "interest in stimulating and aiding the application, to basic biological problems, of the techniques, experimental procedures, and methods of analysis so effectively developed in the physical sciences." (Rees 1987, p. 501). The Foundation followed Weaver's recommendation, and in the 1933 program statement for the Division of Natural Science they formulated a change in the selection procedure for financial support, emphasizing that "interest in the fields play the dominant role in the selection process. Within the fields of interest, selection will continue to be made of leading men and institutions"[16] (Quoted in Rees 1987, p. 502). These funding changes at the Rockefeller Foundation created favorable conditions for scientists like Rashevsky, who was searching for employment opportunities in academia and was interested in bringing the mathematical method to use in biology. For Rashevsky, the move came in 1934 when he joined Chicago University on a 1-year fellowship from the Rockefeller Foundation.

Rasehvsky's vision was to build a mathematical biology on a physico-chemical basis similar to mathematical physics, which he explained in his 1935-paper on "Mathematical Biophysics" in *Nature*:

> Very little attempt has been made to gain an insight into the physico-chemical basis of life, similar to the fundamental insight of the physicist into the intimate details of atomic phenomena. Such an insight is possible only by mathematical analysis; for our experiments do not and cannot reveal those hidden fundamental properties of Nature. It is through mathematical analysis that we must infer, from the wealth of known, relatively coarse facts, to the much finer, not directly accessible fundamentals. (Rashevsky 1935, p. 528).

Rashevsky perceived the role of mathematics as being crucial. Mathematics should function as a "gateway" to the unknown, as a theoretical "microscope" that has the ability to "reveal those hidden fundamental properties of Nature" that are "not directly accessible." This was also a critique of the experimental approach in the field. Rashevsky wanted the theoretical mathematical analysis to come first, so it could guide experiments. Investigations should begin, he argued, with abstract conceptions:

[15] For Rashevsky's career at the University of Chicago and his relationship with the Rockefeller Foundation, see Abraham (2004). For a scientific biography of Rashevsky, see Shmailov (2016).

[16] The Rockefeller Foundation, *President's Review and Annual Report, 1958*.

> This use of abstract conceptions in the beginning is the characteristic of the physico-mathematical method. Violation of this rule, and all attempts to start with actual cases in all their complexity, result in failure. (Rashevsky 1935, p. 528)

His analogy between cells and droplets is part of such an abstract conception. In 1931, he explained it in a paper in *Physics* using the term "model," claiming that:

> Although this phenomenon [division of an oil drop, suspended in water] shows certain similarities to the division of a living cell, no biologist of course will ever assume, that in the actual cell exactly the same phenomenon occur. However the observation of such a "model" justifies the more general assumption, that some kind of variation of the surface tension of the cell, probably of a much more complicated nature, may be responsible for the cell-division. The study of such models becomes especially valuable, if it suggests new experiments with the living organisms or new points of view for their interpretation. (Rashevsky 1931, p. 143–144)

Methodologically, he subscribed to what he called "paper and pencil models," which he explained as follows:

> [...] and a physicist has enough confidence in the results of his calculations, that he does not need actually to build a model, and may satisfy himself by investigating mathematically, whether such a model is possible or not. The value of such "paper and pencil" models is not only as great as that of actual "experimental" models, but in certain respects it is even greater. The mathematical method has a greater range of possibilities, than the experimental one, the latter being often limited by purely technical difficulties. ... Let us therefore make an attempt at studying such "paper and pencil" models of some biological phenomena but let us always keep in mind, that what we study is but a "model," and that no premature claims at any actual "explanations" of the corresponding biological phenomena are as yet justified. (Rashevsky 1931, p. 144)

He constructed an abstract concept of a cell through the analogy to drops. He considered a drop of a liquid *A*, suspended in another liquid *B*, and investigated properties of this system. He analyzed situations where, e.g., one "food substance" *C* diffuses into the drop from outside and gets transformed into *A*. This would create a difference in concentration outside and inside the drop. Discussing surface energy, he argued that if, for "a certain size of the cell the total surface energies of two half-cells is smaller than the surface energy of the whole cell [...] [it] will divide in two spontaneously." On the basis of his analysis, he concluded that "on the several properties which our theoretical drops possess, we see, that first of all they show what may be called "metabolism." [...] Our drop constitutes indeed a "model" of a simplest living organism (Rashevsky 1931, p. 153). He developed the model step by step in a long series of publications. In a publication a year later, he took the free energy of the cell volume into consideration and argued that when the radius of the cell reaches a certain size, the total free energy of the system decreases by a division of the cell, and by invoking the physical principle of free energy he concluded that "a cell must always divide after reaching a certain size." (Rashevsky 1932, p. 395). He calculated the size of the cell and the time of division and concluded that:

> [...] The circumstance that starting with very general thermo-dynamical considerations, and with a minimum of special assumptions we arrive at correct values for the size of the cell and at reasonable values for the time of division, deserves careful attention. ... we are inclined to believe that they actually give us a clue to the understanding of the general cause of any cell division. (Rashevsky 1932, p. 395)

In 1934, he was invited to the second Symposium on Quantitative Biology at the Cold Spring Harbor Laboratory on Long Island, New York. It had been established in 1890 as a marine biology teacher training laboratory. Charles Davenport (1866–1944) became director in 1898, and he began a research program in genetics. In 1924, Reginald G. Harris (1898–1936) took over as director, and he continued the emphasis on research. He started a series of symposia where mathematicians, physicists, chemists, and biologists could come together for a month-long stay over the summer to learn from each other. In the introduction to the proceedings of the second symposium, he advocated for interdisciplinary approaches to biology:

> [...] the more complex the problem, the more the biologist must use mathematics, physics and chemistry, and the more valuable cooperation with representatives of these several sciences becomes. (Harris 1934)

On this account, Rashevsky was in full agreement with Harris; he almost echoed Harris' words in his 1935-paper in *Nature*. Rashevsky presented his thoughts about using the mathematical method in biology to investigate the phenomenon of cell division at the symposium. Both Rashevsky's talk and the discussion that followed were published in the proceedings of the symposium. He presented his ideas about investigating an abstract conception of a cell and claimed in the introduction that cell division could be explained as a direct consequence of the forces arising from cell metabolism. It could be done, he said, "logically and mathematically from a set of well defined general principles," claiming the authority of mathematics:

> It is only natural to assume that the lack of our knowledge of the fundamental causes of biological phenomena, in spite of the tremendous amount of valuable facts, is due to the lack of use of deductive mathematical methods in biology. (Rashevsky 1934, p. 188)

In the analysis, he presented at Cold Spring Harbor he added yet another component to his model—he included investigations at the level of molecules. He derived an expression for the force exerted on each volume of the solvent by the solute, the force action on each volume as a result of osmotic pressure, and the force of repulsion between molecules. The sum of these forces, he argued, is the force per unit volume that a gradient of concentration (caused by metabolism) produces per unit volume. To reach what he called a "general qualitative picture," he calculated the effects of these forces on a homogeneous and spherical cell that is for a further idealization of his abstract conception of a cell. His calculations showed that when the radius of the cell exceeds a certain size then the increase in surface energy caused by a division of the cell is less than the decrease in volume energy. In an earlier paper, this had led him to conclude that:

> As any system tends to assume a configuration, for which its free energy has the smallest value possible, one is tempted to infer that therefore division of a cell will occur spontaneously as soon as, [...] the cell will exceed the critical size. (Rashevsky 1934, p. 192)

But by the time he spoke at Cold Spring Harbor, he had realized that "Unfortunately [...] things are not so simple" so on that occasion he concluded only that:

> every cell, by virtue of the processes of metabolism [...] contains in itself the necessary conditions for spontaneous division above a certain size. (Rashevsky 1934, p. 192)

From the discussion published in the proceedings after Rashevsky's paper, it is clear that the biologists did not approve of Rashevsky's method. They asked critical questions about both the various simplifications, assumptions, and idealizations Rashevsky made in the process of modeling the phenomenon of cell division as well as whether he was modeling "the right thing." For instance they asked, "What example in nature would be nearest to this theoretical case?"

Davenport summarized the general opinion of a typical biologist concerning Rashevsky's approach:

> I think the biologist might find that whereas the explanation of the division of the spherical cell is very satisfactory, yet it doesn't help as a general solution because a spherical cell isn't the commonest form of cell. The biologist knows all the possible conditions of cell form before division [...] There doesn't seem to be in any general way a relationship between the form or size in connection with the cell division. (Rashevsky 1934, p. 197–198)

Rashevsky seemed to have gotten more and more frustrated with the critical reaction from biologists, and in response he tried to explain the method, again with reference to its success in mathematical physics:

> I have insisted on several occasions that the results presented today are only the first steps in the development of mathematical biology. It would mean a misunderstanding of the spirit and methods of mathematical sciences should we attempt to investigate more complex cases without a preliminary study of the simpler ones. [...] To my mind it is already quite a progress, that a general physico-mathematical approach to the fundamental phenomena of cellular growth and division [...], had been shown to be possible. Judging by the development of other mathematical sciences, I would say that it will take at least twenty-five years of work by scores of mathematicians to bring mathematical biology to a stage of development comparable to that of mathematical physics. (Rashevsky 1934, p. 198)

Rashevsky's reasoning rested, on the one hand, on the resemblance between his "paper and pencil model" of an abstract conception of a cell as a drop of a liquid *A*, suspended in another liquid *B*, that metabolizes a "food substance" *C* which diffuses into the drop from outside and gets transformed into *A*, thereby creating a gradient of concentration; and, on the other hand, on agreement of the calculated results for cell size and time span for division with empirical data. In Rashevsky's opinion, he showed that even if his model could not account for cell division for certain, at least it was justified as a possibility worthy enough for further investigations.[17] Biologists could not see the usefulness of an approach that was based on imagination and thought experiments (Gedanken experiments). They were not convinced. In their opinion, here expressed in the words of E. B. Wilson, it was:

> futile to conjure up in the imagination a system of differential equations for the purpose of accounting for facts which are not only very complex, but largely unknown [...] What we require at the present time is more measurement and less theory. (Wilson 1934, p.201)

Their general attitude seemed to be that while mathematics could be of use in the study of ecology and population biology, it could not be used in the study of individuals.

---

[17] In her book *Making Sense of Life*, Evelyne Fox Keller (2002, p. 96) evaluated this to be one of the major problems, "a clash in scientific culture" between Rashevsky and the biologists.

This episode from the early development of mathematical biology illustrates certain conditions for the development of interdisciplinary research using mathematics in biology in the 1930s. Some of these are epistemological. Differences in scientific practices and differences in what is considered to be useful knowledge across the disciplines emerge as constraints. There was a clash in practice across disciplinary boundaries, Rashevsky opting for a theoretical approach, criticizing biologists' practice of putting experiments and fact gathering first whereas biologists found that Rashevsky's approach lacked empirical reality and experiments. Other conditions are of a social-economic nature, such as the emphasis on interdisciplinary research at the Rockefeller Foundation whose financial support paved the way for Rashevsky's project at the University of Chicago. His move to Chicago was supported by Louis L. Thurstone, who was chairman at the department for psychology with which Rashevsky was first associated when he moved to Chicago.

Rashevsky was very active in promoting mathematical biology as a research field in academia. He gathered a group of people around him who pioneered the field, including Herbert D. Landahl, who became Rashevsky's first doctoral student and president of the Society of Mathematical Biology in 1981. Rashevsky also started the scientific journal *The Bulletin of Mathematical Biophysics* as an outlet for their research. It was the first journal in the field. It has been renamed *Bulletin of Mathematical Biology* which is now the official journal for the Society of Mathematical Biology. In the three editions of his monograph *Mathematical Biophysics: Physicomathematical Foundations of Biology*, Rashevky and his group's work can be followed. The first edition was published in 1938 and consists of works on cells and the nervous system. The second, enlarged edition, appeared in 1948, and the last edition was published in 1960. At that time, the book had grown into a two-volume publication. Over the years, Rashevsky began to look into the possibilities of formulating general mathematical principles of biology.

3.3 The Significance of Teaching and Education

One aspect that is clearly present in both episodes as an important factor for the development is the availability of funding. The involvement of civilian scientists in the war effort in the USA and the flow of military and government support for research that followed in the post-war period created favorable conditions and opportunities for mathematicians to become engaged in projects in linear and non-linear programming, game theory, and related areas. Hopes for biology and an emphasis on bringing "some of the methods and machines that had been so successful in the physical sciences," as Weaver phrased it, including mathematics, to the life sciences created employment opportunities and funding for Rashevsky and his project of establishing a group of researchers to develop mathematical biology.

Another aspect is the significance of teaching and study programs for securing a (sub)discipline and developing a body of knowledge—an aspect that the actors were concerned with in both episodes. With respect to mathematical programming, Fred

Rigby, the head of the Office of Naval Research's logistic branch, gave the following evaluation of the state of affairs of mathematical programming in a letter to Mina Rees:

> One [trend affecting mathematics in universities] is for the mathematical aspects of optimisation to find homes in the application disciplines and to be neglected, or at least little respected, in mathematics departments. (From Rigby to Rees (1977, p. 111–112)).

Mathematical programming became attached to operation research (OR) in the post-war period. Operation Research (or Operations Analysis) originated in England in connection with the development of radar to detect aircraft. It represented research into the operational side of the detection system as opposed to more technical sides. During the war, operation research groups consisting of a mixture of various kinds of scientists (e.g., physicists, mathematicians, biologists, chemists, engineers) who analyzed operations, suggested improvements, and worked on strategy and logistics among other things (Fortun and Schweber 1993, p. 602)[18]. In the USA, the Applied Mathematics Panel under the leadership of Warren Weaver developed their own training programs in OR in 1943 (Owens 1989). One of the most famous OR group in the USA during the war was the Antisubmarine Warfare Operations Research Group (ASWORG) headed by the MIT physicist Philip Morse (1903–1985).

There was a strong effort to develop OR for peacetime purposes where education and teaching were seen as key components. OR entered academia in the USA through the military-university complex. Office of Naval Research and the National Research Council funded OR-activities at the universities. Morse was a key player in shaping OR for academia. The first two courses in OR at MIT were held at the mathematics department in 1948. The course report stated that

> a strong effort was made to convey certain basic principles and viewpoints of operations research. [...] Real problems do not fall neatly into one or another of the usual categories of knowledge, but cut widely across boundaries. Successful O/R [requires that the] wholeness of the problem [was] not artificially suppressed. (Quoted in Thomas (2012, p. 110))

This approach was also reflected in the content of the course where there were lessons on formulation and solution of real problems, as well as subject matter such as statistics, search theory, and game theory. The course report also reflects a concern about the behavior of the students, especially those majoring in mathematics who had a tendency to bypass the openness of the problems and turned them into problems in pure mathematics (Thomas 2012, p. 111). Promoting a certain practice through the teaching of courses was an explicit concern. The strong ties to mathematic were reflected in Morse's thinking about training in OR which he thought of as a new field of applied mathematics (Morse 1948, p. 621).

Morse viewed teaching and study programs as vital for securing the further development of OR, the mathematics of OR with linear programming as a core subject, and its establishment in the body of mathematical knowledge. In his speech "Where is the new blood?" at the dinner of the seventh national meeting of the soci-

[18] See also Rau (1999).

ety of OR in Los Angeles in 1955 he expressed the urgency of training and education as he saw it:

> Operations research, as a unified discipline, is just starting. We have just begun to develop our own theoretical and experimental techniques and we have not taught these techniques to very many people yet. The danger-as I see it in the next few years-is that our first few successes will tempt us to go after more successes rather than to work towards a healthy, all-round growth, in fundamental research and in training as well as in immediate results.
>
> Our work in basic theory is coming along slowly, our development of experimental techniques has been less satisfactory. Least satisfactory of all, however, is our record in recruiting and training. I believe we have, up to now, completely failed to train up an adequate supply of new workers in operations research. 'Train' is not the right word though –'recruit' is a more descriptive one. We are failing to attract enough bright youngsters with a flair for scientific research. We are attracting so few that, in the future, our work will not even continue at its present level, let alone expand as it should. The recruiting should not be too hard, either. We don't yet need to catch them at high school age. (Morse 1955, p. 386).

The importance and concern that the historical actors attached to education and teaching for the development of their (new) field are also very much present in the case of mathematical biology. However, the two situations differ from one another. In the first episode of mathematical programming, some of the leading actors attached the new field of inquiry of OR to the well-established discipline of mathematics—Morse called OR a field of applied mathematics and the newly developed linear and more general mathematical programming the core of OR. In contrast, the situation was different for mathematical biology where one established discipline tried to migrate into another established discipline in order to change scientific practice. Reginald G. Harris, the director of Cold Spring Harbor, discussed the value of mathematics in biology in *Scientific Monthly* in 1935 in positive but cautious terms:

> [...] it appears to me that one may expect sufficiently valuable returns from a theoretical biology, based on mathematics, to justify its birth and controlled nurture; this in spite of the fact that there are plenty of examples of the failure of such a procedure in the past (Harris 1935, p. 508)

He referred both to Davenport's conclusion that Rashevsky's work on cell division did not help as a general solution, and to Rashevsky's statement at the Cold Spring Harbor symposium that it would take at least 25 years of work by scores of mathematicians. Harris became quite specific about how one could support theoretical biology while waiting.

> Meanwhile it would seem fair to conclude that theoretical biology, in the sense outlined, should receive some attention as a definite part of biology. I would suggest that half a dozen chairs for theoretical biologists be established at biological laboratories. (Harris 1935, p. 509)

Three of these chairs, he recommended could be in the USA, one in England, and two on the continent. Furthermore, he recommended that they be placed at research institutions and *not* at universities, due to what he saw as an unfavorable consequence, the "dangers" of teaching and education:

> It would seem preferable to establish such chairs at research institutions, in so far as is feasible. If, however, some are established in universities, it should be clearly understood that courses should not be given in theoretical biology. This should be understood for several reasons. The holder of the chair should have as much time as he can possibly use for study and deduction. Furthermore, a professor must be uninspiring indeed who, regardless of his desires, does not, by his teaching alone, beget a number of mental sons and daughters. There are many reasons to believe that we do not wish a flock of newly hatched and hatching theoretical biologists at this time or within the next ten years at least. What we wish is half a dozen brilliant minds to further explore the possibilities of theoretical biology, and to be in a position to become the chiefs of staff if and when recruits are needed. (Harris 1935, p. 509–510).

Because, he explained:

> Many a good thing has been run into the ground because too many hounds followed it too closely and gave to much voice to the chase. (Harris 1935, p. 510).

However, this was exactly what Rashevsky wanted in order to secure the development of mathematical biology. He managed to get a position at the University of Chicago, where he established a section on mathematical biophysics for him and his students in the department of physiology during the 1940s. He then became the chair of an independent committee on Mathematical Biology that oversaw the first degree-granting program in the field in Chicago. During the 1950s, the committee faced a drastic cut in its budget and Rashevsky was unable to keep up the funding from the Rockefeller Foundation. The budget was restored in the 1960s and the committee expanded.

Education and training were issues in securing further developments and the establishment of mathematical biology. In 1961 around 100 people gathered at Western Carolina University to join the Cullowhee Conference on Training in Biomathematics. In the proceedings of the conference the discussion between the participants can be followed. They are at times quite outspoken about their agenda, using vivid analogies about the role and the importance of teaching and education programs in biomathematics. During their discussion of training, for example, the chairman R. G. Miller from Stanford University asked what it would "take to get such a program off the ground?" and he continued:

> If you will excuse me the analogy, I think we have to switch our gaze from philosophical outer space down to the launching pad, and the purpose of this session is to try to indicate how we go about lifting biomathematics into orbit in the university and the scientific community. (CCTB 1962, p. 351)

Also A. F. Bartholomay from Harvard Medical School saw training and education as the tools to transform "biomathematics from art to science." He also emphasized that a common understanding across the disciplines was paramount for the success of this endeavor:

> Thus we must take care that the discipline being created will encourage bold new thinking, yet, at the same time, it must be realized that survival and perpetuation of that discipline will depend on its acceptance into the general scientific environment, and on its capacity to yield results that are meaningful to the general understanding. (CCTB 1962, p. 367)

4 Conclusion

The historical investigations of the two episodes established a point of departure in the protagonists' concrete "projects" and activities in and with mathematics, taken into account the actors' plans, motivations, and ambitions. These were situated in the actors' present contexts which, in turn, were conditioned by various kinds of circumstances from the past and modified by the actors' present opportunities and constraints of various kinds: historical, social, economic, institutional, mathematical, geographical, educational, and so forth, as well as their expectations for the future, see Fig. 6.5. In this sense, the analyses have been exploring the interplay between developments of mathematics and conditions of its developments.

The analyses were done from various perspectives. Internal mathematical ones of the conditions for how specific connections between two areas of mathematics were made. External sociological conditions provided by the military-university complex in the USA during and after World War II. The combination of these two perspectives in the analyses shed light on the interplay between developments of mathematics and conditions in society. The military-university complex in the USA generated funding for research in mathematics, and created channels of communication between academic institutions and the military, and between mathematicians working in academia and mathematicians working on specific practical problems in the military. Together, these interactions paved the way for mathematical programming to develop into a new mathematical research area in academia. The second episode was studied from the perspectives of how protagonists worked with, argued for, and viewed the epistemic role of mathematics in biology. This episode highlighted the conditions for interdisciplinary research including funding and differences in the perception of what constitutes useful knowledge across disciplinary boundaries and different scientific practices at the time. A multiple perspective approach to history of mathematics opens a multiplicity of "windows" through which historical episodes can be investigated and explored, most likely pointing

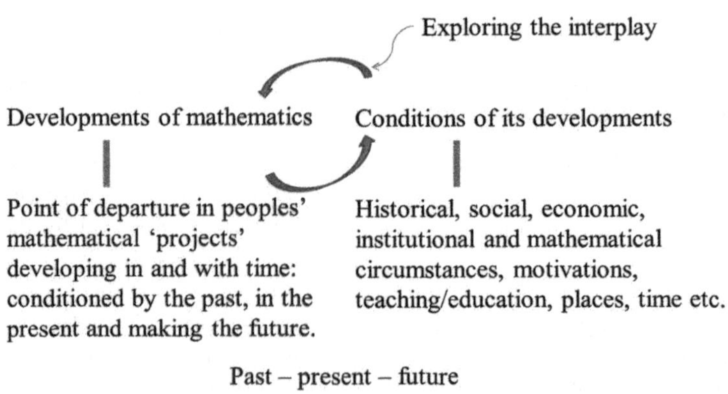

Fig. 6.5 Interaction between developments of mathematics and conditions of its development

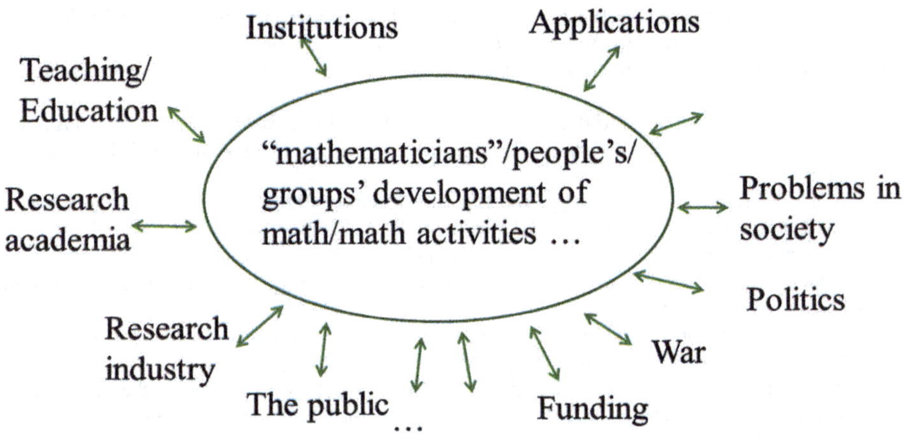

Fig. 6.6 A sketch of a multiple perspective approach to history of mathematics

towards different kinds of decisive factors acting together in a web of events that had an influence on developments of mathematics, see Fig. 6.6.

References

Abraham, Tara. 2004. Nicholas Rashevsky's mathematical biophysics. *Journal of the History of Biology* 37: 333–385.

Belhoste, Bruno. 1998. Pour une réévaluation du rôle de l'enseignement dans l'histoire des mathématiques. *Revue d'Histoire Des Mathématiques* 4: 289–304.

CCTB. 1962. *The Cullowhee Conference on Training in Biomathematics*. Raleigh: North Carolina State University.

Dantzig, George B. 1963. *Linear Programming and Extensions*. Princeton, NJ: Princeton University Press.

———. 1982. Reminiscences about the origins of linear programming. *Operations Research Letters* 1: 43–48.

———. 1991. Linear programming. In *History of Mathematical Programming, A Collection of Personal Reminiscences*, ed. Jan Karel Lenstra, Alexander H.G. Rinnooy Kan, and Alexander Schrijver, 19–31. Amsterdam: North-Holland.

Dantzig, George B., and Marshall K. Wood. 1949. Programming of interdependent activities, I general discussion. *Econometrica* 17: 193–199.

Epple, Moritz. 2000. Genies, Ideen, Institutionen, mathematische Werkstätten: Formen der Mathematikgeschichte. *Mathematische Semesterberichte* 47: 131–163.

Farkas, Julius. 1895. Über die Anwendung des mechanischen Princips von Fourier. *Mathematische und Naturwissensschaftliche Berichte aus Ungarn* 12: 263–281.

———. 1897. Die algebraischen Grundlagen der Anwendung des Fourier'sschen Principle in der mechanik. *Mathematische und Naturwissensschaftliche Berichte aus Ungarn* 15: 25–40.

———. 1899. Die algebraischen Grundlagen der Anwendung des mechanischen Princips von Fourier. *Mathematische und Naturwissensschaftliche Berichte aus Ungarn* 16: 154–157.

———. 1901. Theorie der einfachen Ungleichungen. *Journal für Die Reine and Angewandte Mathematik* 124: 1–27.

Fortun, Michael, and Sam S. Schweber. 1993. Scientists and the legacy of world war II: The case of operations research (OR). *Social Studies of Science* 23: 595–642.

Fréchet, Maurice. 1953. Commentary on the three notes of Emile Borel. *Econometrica* 21: 118–124.

Harris, Reginald G. 1934. Introduction to volume II. In *Cold Spring Harbor Symposia on Quantitative Biology*, ed. Harris G. Reginald, vol. 2, xi–xii. Long Island, NY: Cold Spring Harbor.

———. 1935. Mathematics in biology. *The Scientific Monthly* 40 (6): 504–510.

Heims, S.J. 1980. *John von Neumann and Norbert Wiener*. Cambridge, MA: The MIT Press.

Jensen, Bernard E. 2003. *Historie—Livsverden og Fag*. Copenhagen: Gyldendal.

Kantorovich, L.V. 1960. Mathematical methods of organizing and planning production. *Management Science* 6: 366–422.

Keller, Evelyne Fox. 2002. *Making Sense of Life: Explaining Biological Development with Models, Metaphors, and Machines*. Cambridge, MA: Harvard University Press.

Kjeldsen, Tinne H. 2000. A contextualized historical analysis of the Kuhn-Tucker theorem in nonlinear programming: The impact of world war II. *Historia Mathematica* 27: 331–361.

———. 2001. John von Neumann's conception of the minimax theorem: A journey through different mathematical contexts. *Archive for History of Exact Sciences* 56: 39–68.

———. 2009. Abstraction and application: New contexts, interpretations in twentieth-century mathematics. In *The Oxford Handbook of the History of Mathematics*, ed. Elenor Robson and Jackie Stedall, 755–778. New York: Oxford University Press.

———. 2012. Uses of history for the learning of and about mathematics: Towards a theoretical framework for integrating history of mathematics in mathematics education. In *Proceedings of the International Conference on History and Pedagogy of Mathematics (HPM): 1-21. The HPM Satellite Meeting of ICME-12, July*, 16–20. Daejeon: Korean Society of Mathematical Education & Korean Society for History of Mathematics.

———. 2017. An early debate in mathematical biology and its value for teaching: Rashevsky's 1934 paper on cell division. *The Mathematical Intelligencer* 39 (2): 36–45.

Kjeldsen, Tinne H., and Morten Blomhøj. 2013. Developing students' reflections about the function and status of mathematical Modeling in different scientific practices: History as a provider of cases. *Science & Education* 22 (9): 2157–2171.

Kuhn, Harold W. 1976. Nonlinear programming: A historical view. *SIAM-AMS Proceedings* 9: 1–26.

Kuhn, Harold W., and Albert W. Tucker. 1950. Nonlinear programming. In *Proceedings of the Second Berkeley Symposium on Mathematical Statistics and Probability*, 481–492.

———. 1958. John von Neumann's work in the theory of games and mathematical economics. *Bulletin of the American Mathematical Society* 64: 100–122.

Leifman, L.J. 1990. In *Functional Analysis, Otimization, and Mathematical Economics: A Collection of Papers Dedicated to the Memory of Leonid Vital'evich Kantorovic*, ed. L.J. Leifman. Oxford: Oxford University Press.

Mørch, Søren. 2010. *Store Forandringer*. Copenhagen: Politikens Forlag.

Morse, Philip M. 1948. Mathematical problems in operations research. *Bulletin of the American Mathematical Society* 54: 602–621.

———. 1955. Where is the new blood? *Journal of the Operations Research Society of America* 3: 383–387.

Owens, L. 1989. Mathematicians at war: Warren weaver and the applied mathematics panel, 1942-1945. In *The History of Modern Mathematics, II: Institutions and Applications*, ed. David Rowe and John McCleary, 287–305. San Diego: Academic Press, Inc.

Rasehevsky, Nicolas. 1932. Contributions to the theoretical physics of the cell. *Protoplasma* 16: 387–395.

Rashevsky, Nicolas. 1928. Zur Theorie der spontanen Teilung von mikroskopischen Tropfen. *Zeitschrift für Physik* 46: 568–593.

———. 1931. Some theoretical aspects of the biological applications of physics of disperse systems. *Physics* 1: 143–153.

———. 1934. Physico-mathematical aspects of cellular multiplication and development. *Cold Spring Harbor Symposia on Quantitative Biology* 2: 188–198.

———. 1935. Mathematical Biophysics. *Nature* 135: 528–530.

Rau, Erik Peter. 1999. *Combat scientists: The emergence of operations research in the United States during World War II. Dissertations available from ProQuest*. AAI9926187. https://repository.upenn.edu/dissertations/AAI9926187

Rees, Mina S. 1977. Mathematics and the government: The post-war years as augury of the future. In *The Bicentennial Tribute to American Mathematics, 1776-1976*, ed. D. Tarwater, 101–116. Buffalo, NY: The Mathematical Association of America.

———. 1987. Warren Weaver, July 17, 1894-November 24, 1978. *Biographical Memoirs, National Academy of Sciences of the United States of America* 57: 492–530.

Rellstab, U. 1992. New insights into the collaboration between John von Neumann and Oskar Morgenstern on the *Theory of Games and Economic Behavior*. In *Towards a History of Game Theory*, ed. E. Roy Weintraub, 77–94. Durham and London: Duke University Press.

Richards, Joan L. 1995. The history of mathematics and 'L'esprit humain': A critical reappraissal. *Osiris* 10: 122–135.

Schubring, Gert. 2001. Production Mathématique, Enseignement et Communication. Remarques sur la note de Bruno Belhoste, "Pour une réévaluation du rôle de l'enseignement dans l'histoire des mathématiques" parue dans la RHM 4 (1998), p. 289–304. *Revue d'Histoire Des Mathématiques* 7: 295–305.

Shmailov, Maya M. 2016. *Intellectual Pursuits of Nicolas Rashevsky. The Queer Duck of Biology*. Basel: Birkhäuser, Springer International.

Thomas, William. 2012. Operations research vis-à-vis management at Arthur D. Little and the Massachusetts Institute of Technology in the 1950s. *Business History Review* 86: 99–122.

von Neumann, John. 1928. Zur Theorie der Gesellschaftsspiele. *Mathematische Annalen* 100: 295–320.

———. 1937. Über eine ökonomische Gleichungssystem und eine Verallgemeinerung des Brouwerschen Fixpunktsatzes. In *Ergebnisse eines Mathematischen Kolloquiums Wien*, ed. Karl Menger, 73–83.

von Neumann, John, and Oskar Morgenstern. 1944. *Theory of Games and Economic Behavior*. Princeton: Princeton University Press.

von Neumann, J., and Fréchet, M. 1953. Communication on the Borel Notes. *Econometrica* 21: 124–127.

Wilson, E.B. 1934. Mathematics of growth. In *Cold Spring Harbor Symposia on Quantitative Biology*, vol. 2, 199–202. Long Island, NY: Cold Spring Harbor.

Chapter 7
Teaching as an Indicator of Mathematical Practices

Carlos Tomei

Abstract In their study of the relationship between teaching and mathematical research, Belhoste and Schubring introduce a theoretical framework and a wealth of interesting case studies, highly substantiated by their study of primary sources. Following on their steps, I take teaching as a fertile indicator of the community's varying points of view about its practices, ranging from the interaction with other disciplines to the development of the strict mathematical corpus. As suggested by Schubring, an interdisciplinary approach is requested for a better evaluation of the influence of mathematics on its practitioners and with other fields of knowledge. Some lines of inquiry are presented.

Keywords Teaching · Learning · Mathematical practices · Epistemology · Textbooks

1 Introduction

As a working mathematician, I frequently compare myself as a keeper of a wonderful tradition. After millennia of contributions, mathematics is a diverse, powerful, critical body of knowledge. Its commentators add to the overall feeling of awe and curiosity. Philosophers ask about the role of logic in mathematics and in mathematicians' practices. Researchers in cultural studies consider its pervasive effect from pure science to the common man's attitude with respect to quantification and structure. The reciprocal influence among internal modalities—pure and applied mathematics, teaching, and research—provides a fertile ground for epistemological debate: any aspect seems open ended.

It is thus not surprising that authors have different opinions about the interaction between teaching and the development of mathematical theory. I take as a starting point for my comments as a practitioner the juxtaposition of some of Belhoste's and

C. Tomei (✉)
Departamento de Matemática, PUC-Rio, Rio de Janeiro, Brazil
e-mail: tomei@mat.puc-rio.br

G. Schubring (ed.), *Interfaces between Mathematical Practices and Mathematical Education*, International Studies in the History of Mathematics and its Teaching,
https://doi.org/10.1007/978-3-030-01617-3_7

Schubring's points of view. Belhoste's (1998) original article was reviewed by Schubring (2001), which then extends on the matter in a subsequent text (Schubring 2018). One should not expect in this essay the erudition of both authors, as displayed in their books (Belhoste 1995; Schubring 2005).[1]

Following Kuyk (1977; see Schubring 2001, p. 297), Belhoste and Schubring believe that the influence of teaching in the development of mathematical practices is underrated. In particular, Belhoste comments on the indifference towards the role of teaching in the historical development of mathematics. Schubring, in his subsequent articles, is enthusiastic about establishing a program dedicated to the reevaluation of the role of teaching.

A loose interpretation of teaching as the means of communication, transmission, and diffusion of mathematical knowledge may lead us to think that there is little left outside of teaching. Clearly, socialization in different forms is fundamental among researchers, students, and practitioners in general: at the risk of trivializing the issue, one must specify relevant contexts. In a similar vein, there is too much encapsulated under the heading of mathematical research: the historiography of what is considered "doing theory" is one of the most interesting aspects in history of the subject. Belhoste and Schubring agree that it is hard for the historian to separate production from the conditions of reproduction.

A satisfactory common ground to approach this ecology of concepts is achieved by the consideration of case studies, and both authors present expressive examples, to be commented later. In the words of Belhoste, teaching is a form of socializing knowledge, by converting it into normal science. Historically, teaching was fundamental in the creation of a career—the researcher in mathematics—and it was instrumental in the specification of different areas of mathematics through the compartmentalization of courses.

In order to bring specificity to the study of the rather ambiguous practices, we call teaching and research, Schubring suggests a conceptual framework which naturally leads to interdisciplinary investigations. Two examples suffice to show the vastness of the project: the effect of ideology and religion on teaching, the reciprocal influence of practices of different subjects and their effect on society.

2 Textbooks: From Babylonia to Bourbaki

A wealth of very well-chosen examples is presented by Belhoste (1998) and Schubring (2001; this volume). The purpose of my very schematic presentation is to indicate some relevant aspects to be followed along the text.

Schubring comments on Proust's work on Babylonian clay tablets (Proust 2014). She sorts some 2000 pieces and distinguishes among exercises for young scribes,

[1]A *caveat:* Despite the fact of having Turing as one of my intellectual references, I will not refrain from using words as thinking, understanding, and clarity. The inevitable imprecision should be compensated by the brevity of the presentation.

the occasional tables, problems followed by their resolution, and problems which seem to be open, expecting consideration. The whole gamut of practitioners is covered, from beginning students to scholars. Jumping to the nineteenth century, Schubring quotes an observation of Desttut de Tracy in his *Éléments D'Idéologie* (1801): the preparation of textbooks requires creativity for the appropriate presentation of known facts, a kind of research by itself. Cauchy in his years at the École Polytechnique attached research to teaching by establishing levels of rigor, a pedagogical (ideological, according to Schubring) requirement. Dedekind points out to the lack of foundations which led to his *Stetigkeit und irrationale Zahlen* (1872).

Meyer wrote about sets (1885) while Cantor was still perfecting his theory. From (Schubring, this volume),

> [...] Cantor's ideas were for Meyer ... only the trigger for developing fundamental concepts developed already since a long time by mathematics teachers. He did not present the concept of set as something new, but as belonging to a tradition going back to the ancient Greeks.

The debate about teaching through set theory seems to begin here. According to Schubring, Klein was strongly against it: it required further understanding and did not seem fertile. Again, from (Schubring ibid), "... [the subject seemed far] from having matured to the point of having induced and intradisciplinary process of integration and restructuration. The concepts of set theory did non (yet) provide new elements for mathematics."

Schubring acknowledges the importance of Beltrami's text of 1868 to the acceptance of non-Euclidean geometry by the mathematical community as a whole. The concrete model presented in the text is exemplary—how to distinguish among research, clarification, the need to convince the skeptics? Schubring draws attention to another author on the subject, Wagner, which, like Meyer, makes an effort to combine teaching and rigor. Significantly, his book has two prefaces, for colleagues and for students.

Belhoste ascribes the origins of Bourbaki's monumental project to the desire of an alternative text to Goursat's analysis book. In his conclusion, Schubring points out to an interesting issue:

> [...] it is not possible to close our eyes to the fact that dogmatic, formalizing impulses for the development of science have also emerged from the school: in a fundamentalist exaggeration of the search for firm foundations (Schubring, this volume).

All such examples are significant, and new ones will surely come up with the expansion of research on these lines. But instead of searching for common trends, I prefer to consider them as indicators of different contexts.

Another set of examples of how subtle is the line between teaching and research is the presence of mathematical results, which first appear as exam questions. Spivak (1965) points out that Stoke's theorem, originally quoted in a letter from Kelvin to Stokes, is also a question on the Smith's Prize Examination for 1854. Bost and Mestre (1988) present a result by Richelot on a generalization of the arithmetic-geometric-mean for surfaces of genus 2 with a similar fate. One is also reminded of more or less apocriphal stories about (non-Babylonian) students who solve open problems thinking that they were homework assignments. The Fáry-Milnor theorem, according to Milnor himself, is not such an example. But, according to Kuperberg (n.d.),

> […] Dantzig was in graduate school in statistics, it was 1939, he came in late to class, he saw two open problems on the board and thought they were homework. The professor was Neyman, who must have been a bit impenetrable to his students, because he didn't tell Dantzig for six weeks what had really happened.

3 Professionalization

Both Belhoste and Schubring expand on the importance of teaching for the characterization of a mathematical profession, and again the issue has a number of different aspects. Indeed, there is a substantial difference between the importance given to mathematics by those who study music in the *quadrivium* and the fact that the State needs administrative specialists, not to mention statisticians. Cauchy is certainly an interesting case study for the rise of the public servant, but the teaching of applied mathematics was already present in universities.

There is a relevant curve to follow in the historical process: university calculus seems to be disconnecting from the demands of the professional world. Other university departments provide mathematical material, which correlates better with a computerized world of large scale modeling and simulation. Is there a difference in what is taught between private and public contexts?

Thurston (1994) provides a wonderful case study of the economics of mathematical research. His results in foliation theory were so good that students were suggested to drop the subject. The influence of trends in mathematics looks small when compared to physics.

The necessity of an interdisciplinary point of view is clear: in some situations, mathematics gets stimulated and in others, confined. Schubring considers the employment of mathematical teaching to religious demands, a rich subject. One is tempted to extrapolate: mathematics seems to be more accepted as a tool for higher efficiency in the armies than in other fields of the government.

4 An Interdisciplinary Project

As Schubring points out, the interaction of mathematics with other cultural activities is an interdisciplinary subject: in order to understand mathematical production, one must see it in relation with other social subsystems. The scale of such a project is immense, the influence of teaching encompasses the presence of mathematics in our intellectual life.

The issue is unavoidable and Schubring states it clearly (Schubring 2001): there was never autonomy for mathematics (and here he is emphasizing the teaching activities). In secondary school, mathematics is a concurring subject, together with other values conveyed by ideology. In universities, it negotiates with other disciplines. Research centers before 1800 would concentrate in teaching (and nowadays such institutions require substantial support from the State).

Below, three lines of investigation are barely sketched.

4.1 Mathematics as an International Activity

Schubring proposes as a research subject the confluence of local projects leading to an international mathematical community (Schubring 2001).

Following Luhmann (1990) and Stichweh (1984), he takes communication as the elementary act of science, foundational to teaching, learning, and doing research. Communication requires a common language and a shared culture, and institutions and states become protagonists of the supranational interchange.

The original ideas shaping mathematics are so old that the temptation of treating them as ahistorical is unavoidable. A schoolboy learning Pythagoras' theorem (a beautiful example of incorporation of mathematical knowledge of yet older times) probably has no idea of Greek history, or philosophy.

4.2 Collective and Individual Practices

There is no need for agreement on the matter: interesting examples of both positions are abundant. A recent improvement on mathematical communication is the presence of very active blogs, like Tao's or Mathematics Stack Exchange.

Thurston's presentation (1994) of his own mathematical experience borders on the dramatic. He begins emphasizing the introspection required for his understanding of foliation theory, which included intensive computer programming: "The standard of correctness and completeness necessary to get a computer program to work at all is a couple of orders of magnitude higher than the mathematical community's standard of valid proofs."[2]

Thurston then describes the necessary effort for spreading his knowledge. His statement is an inextricable merging between teaching and research: one does mathematics to obtain understanding and it takes energy to convey not only proofs of new results, but "the ways of thinking about the same mathematical structure." Belhoste (1998) quotes the many contributions to elliptic function theory in the mid-nineteen century as teaching landmarks, and Schubring's opinion (Schubring 2001) is that of a practitioner—this is pretty much what research is about, the natural

[2]The interaction with computers is already a subject for the history of mathematical practices. Different reactions within university departments are leading to a radical design of professional and teaching standards.

development of concepts.[3] Indeed, the community assigns different values to the verification of a claim and its absorption in the corpus.

There is a situation in the boundary between private and public practices which deserves more attention. A common place within the community is how much there is to learn from a teacher who thinks while giving his class (as opposed, say, to someone who just follows a prepared text). There is some truth to it: errors should be part of teaching, there are few occasions to appreciate a specialist in intellectual action. There are also familiar opposing arguments: students (and teacher) may get confused, intimidation may pop up at any time.[4] But perhaps there is more to it: some people might prefer to think in public—a positive form of vanity stimulates the capacity of verbalizing a subject.

In a somewhat reciprocal fashion, one is led to think if genuine collective study would eliminate the case studies referred in Sokal and Bricmont (1997).

4.3 *Abstracting, Forgetting, Getting to the Point*

A prehistoric man who owns a few sheep finds a stone which reminds him of one of them. At some point, he may have gathered a collection of such amulets, representations of the animals, which are so dear to him. It must have taken a long time for a collection of rocks to become an indicator of the size of a flock, possibly the first application of a forgetful functor. One wonders: how many squares have to be drawn side by side in a vase until *the* square becomes *a* square? Coherently, the usual references to the origins of counting do not treat these issues.

Abstracting is not natural: it took a figure of authority like Hilbert to make the mathematical community get used to the fact that some initial words are devoid of meaning when building up an axiomatic system, and this certainly helped the acceptance of non-Euclidean geometries, as observed by Gray (1994). Why is a circle not a straight line for a topologist?[5] Why are R^2 and R^3 different? One understands that geometry was a good subject to practice argumentation (Schubring 2005), being less polemical than religion or politics, but there is more: the discussion may end. Some intellectual practices thrive on open-endedness: forgetting may become suspicious, one is responsible for whatever is being discarded.

[3] One is reminded of Besicovitch, "a mathematician's reputation rests on the number of bad proofs he has given."

[4] Recall the familiar story attributed not only to some mathematicians, but also to physicists. The teacher states, "This is trivial" and a student asks why. The teacher leaves the room, comes back and makes it clear, "yes, it's trivial," and goes on with his class.

[5] And still "you must remember this... the plane is not a disk..." (As z goes by).

5 Missing Opportunities in Teaching

Why is it that in mathematics we do not acknowledge epistemological bangs, but just the occasional whimper of increasing consensus?[6] Some exhaustively studied episodes in the history of mathematics deserve more recognition (percolation seems to be the right word). The issue affects the contents of our courses and provides indication of a certain hierarchy of relevance, in which conceptual aspects are less important than more pragmatic practices.

Clearly some fundamental problems have ceased to be so: the community, from students to researchers, do not seem to worry anymore about the distinction between numbers and quantities, which is usually considered as being resolved by Stevin (Malet 2006; Schubring 2001), or by the omnipresent rule of signs (Glaeser 1983). I present a couple of issues which may still intrigue a current practitioner.

5.1 Optimal Design

Most calculus students are confronted with the computation of the path of light joining source and target hitting a mirror along the way. What is originally a simple problem in finding a minimum opens up to the metaphysical issues associated to Maupertuis' principle: very informally, nature behaves in the best possible way. I was impacted when entering university by this spectacular evidence of intelligent design much before the spreading of the creationist debate. Teachers both in mathematics and physics did not come up with satisfactory answers. Years later, I learned that I was in great historical company: intellectual comfort came by accident, through Ekeland's wonderful book (2006). To say the very least, we miss an opportunity of presenting one of the great concepts of physics—the idea of averaging over all paths, also known as the Feynman integrals (1985).

And then things turn upside down: instead of brilliant photons, they are so stupid (or homogeneous in their behavior) that it is the appropriate average of all their choices which makes one think that an optimal decision was taken. The point of view is well known in social sciences under the catchphrase "the wisdom of the crowds" (Surowiecki 2004). In optimization and signal processing, the (hard) search for optimal behavior is frequently replaced by sampling a typical object, which sometimes performs in an almost optimal fashion.

[6] Azzouni (2006) refers to the benign fixation of mathematical practice, which he takes to be an almost unique attitude towards consensus, specific to the mathematical community.

5.2 *Instantaneous Interaction*

Consider the criticism against Newton for making use of instantaneous gravitational interaction. Some of his contemporaries filled the universe with stuff that would somehow transmit the forces. Newton discarded the issue, defending his position by simply presenting the output of his calculations. Why shouldn't a contemporary student be less intrigued than Descartes? What is offered in return for this curiosity? Riemann's ideas, which led to what Wheeler would call geometrodynamics—geometry induces physics—are a form of esoteric knowledge, a privilege of few. As far as intellectual stimulation goes, almost nothing compares to the idea that it is the geometry of some enlarged space which makes you think that a planet is pulling down an apple.

5.3 *Method*

The flatness of the Earth is out of fashion: we have seen the photos of a round planet, with the fundamental implication that we live in a bounded surface with no boundaries. Still, for most people, a bounded universe would require a wall enclosing it. Concepts think for us. In this case, we should get used to three-dimensional compact manifolds. And then, one might wonder, what about orientation? I do not think that teaching does a good job in making explicit the power of well-chosen concepts.

Computers generated an abundance of theorems in plane geometry, which seem to be irrelevant to what is considered the body of knowledge on the subject. Indeed, the presentation of a mathematical subject is hardly a thick description, in the anthropological sense. It is more like the description of a city by first indicating its main spots, from which one may provide more detailed information in order to reach any other point. Mathematicians frequently talk about geometrizing mathematics: Euclid's axioms are taken to higher dimensions, to discrete structures. But when we tell students that two distinct points determine a unique line in 200 dimensions, we should make it clear: this is not a visual fact—or is it? We are missing epistemological labs.

Learning is not simple, and the history of mathematics is a collection of impressive difficulties. One of the mantras among practitioners is the fact that the Greeks came up with the axiomatic method as a fertile method for the presentation and generation of mathematics. There is much to learn from verifying the statement throughout the millennia, but also to understand why it is so strongly believed. I outline a few examples.

5.4 Negative Numbers and Virtuality

One could hardly start from a better place than Glaeser's *Épistemologie des nombres rélatifs* (1983) for a presentation of the history of the rule of signs (the product of two negative numbers is a positive number). For a practitioner, the changes of opinion among major mathematicians provide a strong argument for knowing more about the history of the subject. The issue seems to have been resolved satisfactorily to our current standards in a text by Hankel's (1867) on complex numbers: according to Glaeser, Hankel performs the passage from the concrete to the formal level. But there is more, a problem which is presented almost as an understatement by the author:

> La révolution accomplie par Hankel consiste à aborder le problème dans une toute autre perspective. Il ne s'agit plus de déterrer dans la Nature des exemples pratiques qui 'expliquent' les nombres relatifs sur le mode métaphorique. Ces nombres ne sont plus découverts, mais inventés, imaginés.

On the one hand, there was no satisfactory metaphor: what is the product of two debts, one would react at a naive proposal? On the other hand, there was no answer from the Heavens. At this point, the standard practitioner should be intrigued—the axiomatic method has been disregarded for two millennia.

There are psychological aspects behind the change of attitude. Alas, it is not Nature which validates mathematics: validation follows from the axioms which we choose, rules devoid of any transcendence. The virtual world—Turing's paradise—should be the stomping ground of the mathematical community. Hardy should get used to the idea that intelligence services live out of his research.

We are reminded of Rota's description of the double life of mathematics (Rota 1991):

> [...] The facts of mathematics are as useful as the facts of any other science. No matter how abstruse they may first seem, sooner or later they find their way back to practical applications... Axiomatic exposition is indispensable in mathematics because the facts of mathematics, unlike the facts of physics, are not amenable to experimental verification."

5.5 Non-Euclidean Geometry

Few practitioners realize that there was no clear agreement to what a non-Euclidean geometry meant. Gray's text is a short, incisive account (1994), presenting the difficulties and the required change of attitudes. Spherical geometry was out: two lines should not enclose a bounded area. Because of Kant's influence, the right axioms had to fit with our innate conception of space. This is in sharp contrast to our current practice: different geometries simply depend on different starting points. Gray stresses the importance of symbolic language: "...by casting geometry into trigonometrical formulas, Bolyai and Lobachevsky both produced a language for analyzing the behavior of lines that evades the weight of tradition."

For a layperson, the underlying conceptual revolution is surprisingly recent: "The far more radical step of denying mathematical terms any meaning and relying completely on formal rules of inference was taken by Peano and, independently by Hilbert."

The fertility of the new, looser interpretation of the undefined concepts was evident: one could do geometry in so many different contexts, for example, ranging from finite fields to Banach spaces.

5.6 No-Go Theorems, Fertile Imprecision

Some important ideas in physics are prohibitions: there are no monopoles, there is nothing faster than light, certain properties are conserved. Are they theorems? Not really, in the sense that few practitioners would know the underlying hypotheses of such claims. Perhaps a better contextualization would be saying that such statements are true in the light of all the facts we believe.

From a mathematical point of view, the attitude is an acknowledgement of coexisting sets of axioms. This happens also among practicing mathematicians: the Cayley-Hamilton theorem for complex matrices may be proved using Cauchy's theorem for analytic functions, and with a little syntactic push reminiscent of model theory, the argument extends to commutative rings with unity.

Moreover, a physicist is willing to occasionally change the axioms while research advances. An annoying corollary arises in occasional interactions between mathematicians and physicists. The example is taken from Schwartz (1992). The mathematician presents his results on a specific issue (say, the Schrödinger equation for the hydrogen atom) and the physicist immediately asks what happens in a slightly different formulation. Indeed, the equation is a model, built up from hypotheses and simplification and is thus amenable to fudging. Still, an understanding of simple cases is one of the best contributions a mathematician can offer.

5.7 New Students, New Teachers

The idea of what constitutes an elementary subject is an interesting example. One may think that the elementary material is that which is taught to beginners. Logicians and participants in mathematical Olympiads see it differently: a more elementary subject has less pre-requisites. Anachronistically, researchers like Frege and Boole seem like unconsciously preparing material for the dumbest student, the computer. Perhaps the community will soon be talking about machine teaching hand in hand with machine learning.[7]

---

[7] At the risk of being criticized for specism.

6 A Personal Recollection

Perhaps the hardest part in the study of the ties between teaching and research practices is the identification of soft fossils, the many episodes which are not documented in a stable fashion.

The following Zen story is traditional. A monk shows a fan to three students and asks, "what is this?" The first replies, "it's a fan," and the monk is not pleased. He shows the fan to the second student, which picks the fan from the master's hands and starts waving himself. The master is still not satisfied and shows the fan to the third student, which takes the fan and uses it to scratch his back.

This story made more sense to me during my Ph.D. years at the Courant Institute. As a student, I was convinced that powerful mathematics came from the knowledge of better theorems. This changed after being exposed to a number of situations in which the masters would scratch their backs with theorems I knew. After a few examples, I felt entitled, more, I felt invited to do the same. This exercise of intellectual freedom was the most important lesson of my graduate years. Students are surprised when they realize that there are many addition tables, many metrics, all triangles have a straight angle.

Acknowledgements The author thanks Laura Nery for extended conversations. The support of CNPq and FAPERJ is gratefully acknowledged.

References

Azzouni, Jody. 2006. How and why mathematics is unique as a social practice. In *18 Unconventional Essays on the Nature of Mathematics*, ed. Reuben Hersh, 201–219. New York: Springer.

Belhoste, Bruno. 1995. *Les Sciences dans L'Enseigenement Secondaire Français, Tome 1: 1789–1914*. Paris: Institut National de Recherche Pédagogique et Éditions Économica.

———. 1998. Pour une réévaluation du rôle de l'enseignement dans l'histoire des mathématiques. *Revue d'Histoire des Mathématiques* 4: 289–304.

Bost, Jean-Benoît, and Jean-François Mestre. 1988. Moyenne arithmético-géométrique et périodes des courbes de genre 1 et 2. *The Mathematical Gazette* 38: 36–64.

Ekeland, Ivar. 2006. *The Best of All Possible Worlds: Mathematics and Destiny*. Chicago: The University of Chicago Press.

Feynman, Richard. 1985. *QED: The Strange Theory of Light and Matter*. Princeton, NJ: Princeton University Press.

Glaeser, Georges. 1983. Épistemologie des nombres relatifs. *Recherches en Didactique des Mathématiques* 2 (3): 303–346.

Gray, Jeremy John. 1994. Euclidean and non-Euclidean geometry. In *Companion Encyclopedia of the History and Philosophy of the Mathematical Sciences*, ed. Ivor Grattan-Guinness, 877–886. London: Routledge.

Hersh, Reuben. 2006. In 18 Unconventional Essays on the Nature of Mathematics, ed. Reuben Hersh. New York: Springer.

Hankel, Herman. 1867. *Théorie des complexen Zahlsysteme*. Leipzig: Leopold Voss Ed.

Kuperberg, Greg. n.d. *The Story About Milnor Proving the Fáry-Milnor Theorem*. https://mathoverflow.net/questions/54513

Kuyk, Willem. 1977. *Complementarity in Mathematics: A First Introduction to the Foundations of Mathematics and Its History*. Dordrecht: D. Reidel Publishing Company.
Luhmann, Niklas. 1990. *Die Wissenschaften der Gesellschaft*. Frankfurt am Main: Suhrkamp.
Malet, Antoni. 2006. Renaissance notions of number and magnitude. *Historia Mathematica* 33: 63–81.
Proust, Christine. 2014. Does a master always write for his students? Some evidence from old Babylonian Scribal schools. In *Scientific Sources and Teaching Contexts throughout History: Problems and Perspectives*, ed. Alain Bernard and Christine Proust, 69–94. Dordrecht: Springer.
Rota, Gian-Carlo. 1991. The pernicious influence of mathematics upon philosophy. New directions in the philosophy of mathematics. *Synthese* 88 (2): 165–178.
Schubring, Gert. 2001. Production mathématique, enseignement et communication. *Revue d'Histoire des Mathématiques* 7: 295–305.
———. 2005. *Conflicts Between Generalization, Rigor, and Intuition: Number Concepts Underlying the Development of Analysis in 17th-19th Century France and Germany*. New York: Springer.
———. 2018. *The Impact of Teaching Mathematics Upon the Development of Mathematical Practices* (this volume).
Schwartz, Jacob T. 1992. The pernicious influence of mathematics upon science. In *Discrete Thoughts. Essays on Mathematics, Science and Philosophy*, ed. Marc Kac, Gian-Carlo Rota, and Jacob T. Schwartz, 19–26. Boston: Birhauser.
Sokal, Alan, and Jean Bricmont. 1997. *Impostures Intellectuelles*. Paris: Éditons Odile Jacob.
Spivak, Michael. 1965. *Calculus on Manifolds*. Reading, MA: Addison-Wesley.
Stichweh, Rudolf. 1984. *Zur Entstehung des Modernen Systems Wissenschaftlicher Disziplinen: Physik in Deutschland 1740–1890*. Frankfurt am Main: Suhrkamp.
Surowiecki, James. 2004. *The Wisdom of Crowds*. New York: Anchor Books.
Thurston, William. 1994. On proof and progress in mathematics. *Bulletin of the American Mathematical Society* 30 (2): 161–177.

Author Index

G. Schubring (ed.), *Interfaces between Mathematical Practices and Mathematical Education*, International Studies in the History of Mathematics and its Teaching,
https://doi.org/10.1007/978-3-030-01617-3

Subject Index

G. Schubring (ed.), *Interfaces between Mathematical Practices and Mathematical Education*, International Studies in the History of Mathematics and its Teaching,
https://doi.org/10.1007/978-3-030-01617-3

MIX
Papier aus verantwortungsvollen Quellen
Paper from responsible sources
FSC® C105338

Printed by Books on Demand, Germany